Mining Tibet

Mineral exploitation in Tibetan areas of the PRC

Tibet Information Network

London 2002

Printed in England
Published by Tibet Information Network (TIN), November 2002

Design & Typesetting
Jane Bartlett & Matthew Ward

ISBN: 0-9541961-2-0

Mining Tibet

Mineral exploitation in Tibetan areas of the PRC

Images with the copyright 'ART/TOTAR/TOTAR 1997' are from the CD-rom "Tibet Outside the TAR", and have been used with the permission of the authors, Steven D. Marshall and Susette Cooke (of Alliance for Research in Tibet)

Tibet Information Network (TIN):

UK HEAD OFFICE:

City Cloisters,
188-196 Old Street
London EC1V 9FR UK

tel: +44 (0)20 7814 9011
fax: +44 (0)20 7814 9015

US OFFICE:

TIN USA
PO Box 2270
Jackson WY83001

tel: +1 307 733 4670
fax: +1 307 739 2501

tin@tibetinfo.net — www.tibetinfo.net — tinusa@wyoming.com

Contents

Contents continued

Contents continued

Maps

Acknowledgements

Mining Tibet was written by Jane Caple, of Tibet Information Network (TIN), who also carried out the bulk of the research. It would not have been possible for TIN to cover such a complex and wide-ranging subject without the assistance and support of many people. We would particularly like to thank Steven Marshall and Susette Cooke for their guidance and generosity in sharing their knowledge and photographs, also Gabriel Lafitte for his insight and supply of research materials, and a consultant geologist who wishes to remain anonymous but whose technical guidance and insight into the mining industry were invaluable. Many thanks also to the other experts and specialists who gave their time in the midst of very busy schedules, including John Bray, Colina MacDougall, Bradley Rowe, Lorne Stockman and Jude Bueno de Mesquita. Thanks also to Warren Smith, who continues to draw key speeches to our attention.

As with all TIN publications this has been a team effort, largely dependent on the vital contribution of the TIN field researchers in India and Nepal, and also the support of other members of the news and research team in TIN's London office. A huge vote of thanks is due to Jane Bartlett, who freely invested so much of her expertise, energy and time; and generously offered her creative eye and project management skills in the final stages of this project. Credit and thanks go to Jane Hill for her additional work on maps, as well as to Lawrence Crissman and Jason Wotherspoon, who provided the original map outlines. Thanks also to Jess Brownrigg, Tim Kangro and the other volunteers at TIN, who do not wish to be named, for their hard work and enthusiasm. This report would not have been produced without the work and support of former TIN Directors Robert Barnett and Richard Oppenheimer.

The production of this and other TIN reports relies upon the many anonymous individuals who, despite the difficulties and dangers, ensure that information about the situation on the ground continues to emerge from Tibet, and who help to build our understanding of the impact of development on ordinary people. Although unable to publicly acknowledge them, our appreciation and respect goes out to all of these people.

TIN would like to express deep gratitude for the financial support that has contributed to the extensive research behind this report, including that from the Gere Foundation; Loel Guinness and the Kalpa Group; the Staples Trust; the Swedish International Development Cooperation Agency (SIDA); National Endowment for Democracy; the Barrow Cadbury Trust; the American Himalayan Foundation; the Lyndhurst Settlement; the Ruben & Elisabeth Rausing Trust; Schweizer Tibethilfe; The Philanthropic Trust; the Department for International Development Cooperation, Finland; and anonymous benefactors from the UK, France, the Netherlands and the USA.

Preface

> *"In the sand of this wasteland, there are ants by size smaller than a dog but bigger than a fox. (...) While digging their underground galleries these ants bring up auriferous sand. The Indians, therefore, are sent to this wasteland to collect the sand."* Herodotus 3, 102 (4th cent. BC)

It is this legendary account of 'gold-digging ants' in mountainous regions behind the Indus – reported by the Greek historian Herodotus from Persian sources – which probably gave birth to the first myths about Tibet in the Western world. Today's historical science tells us that the account was based on two misunderstandings; due to an erroneous translation, marmots were identified as 'giant ants', and the mountainous 'wasteland' in question is more likely to have been today's Afghanistan than Tibet. Nevertheless, for centuries, the conception of a mythic realm where gold is abundant fired European fantasies about Tibet. Later accounts in the 13th century, like those of William Rubroek and Marco Polo, again reported the existence of fabulous gold reserves in Tibet, although neither of these travellers ever set foot there.

In the following centuries, western travellers to Tibet were mostly Christian missionaries. The reports they left to posterity naturally dealt with their adventurous – and mostly misadventurous – endeavours to spread the Gospel, but few of them failed to mention the fleeting rumours they had heard about gold in Tibet, many invariably adding that there must be more to it than they heard. Remarkably, at this point of time, the western obsession with gold seems to have been unique. In sharp contrast, Chinese sources of the same period only dispassionately mention the existence of gold reserves in Tibet, recording they were only occasionally exploited by the locals.

As one might expect, the strongest Western interest in Tibet's gold and other mineral resources appeared in the 19th century, in the heyday of colonialism. Contemporary travelogues surpass each other with conjectures on the real extent of Tibet's presumed fabulous mineral resources, awaiting the white man's skills and spirit of enterprise to tap them. Andrew Wilson, a Scottish journalist who travelled in the Western periphery of what he called 'Chinese Tibet' (as opposed to the 'Indian Tibet' under direct or indirect British rule – i.e. Ladakh and adjacent areas) and who published the first Himalaya travel bestseller 'The Abode of Snow' in 1875, mentions: "*That there is gold in Chinese Tibet does not admit of a doubt; and in all probability it could be procured there in large quantities were the knowledge and appliance of California and Australia set to work in search of it*".

Such conjectures, however, were based on hearsay rather than real knowledge, since the gate to the 'roof of the world' was to remain closed for most Europeans, until the early 20th century. This very seclusion, in fact, first fired the speculations to the utmost. "*If the richest mineral treasures in the world lie there, as we have so much reason to suppose*" ponders Wilson, "*there is abundant reason why strangers should be kept out of it.*" Wilson finally voices the suspicions of many of his contemporaries by asking: "*Is it possible that the gold – or, to speak more generally, the mineral deposits in Tibet – may have something to do with the extreme anxiety of the Chinese to keep us out of that country?*"

However, the western coloniser's fantasies about exploiting Tibet's minerals were never to materialise. It might be an irony of history that, instead, it was the People's Republic of China, which by supressing Tibet's de facto independence in 1951, embarked upon the first attempts to systematically exploit the mineral riches of Tibet. The first steps to do so date back to the early 1950s, and in the fifty years since then, mineral exploitation has evolved from a rather pioneering and marginal endeavour to a major phenomenon visible in many Tibetan landscapes, with a long-lasting impact on the economy, ecology and many other issues related to Tibet's present and future.

Mining Tibet is the first comprehensive account of this development.

Earth-moving equipment in the vicinity of Mount Kailash **© Max Cotton/TIN 2002**

Introduction

According to the constitution of the People's Republic of China (Article 9) all mineral resources belong to the state, thus from the perspective of the central authorities Tibet's resources are state assets, available to be exploited to meet the needs of national industrialisation, development and revenue generation. The drive to develop the western regions of China, officially launched by President Jiang Zemin in 1999, marked a new push to gain access to Tibet's mineral resources to meet the growing demand for raw materials from Chinese industry, largely concentrated in the eastern coastal provinces.

China's assertion of political control over Tibet in the 1950s opened the door to a completely new economic relationship between Tibet and China. For the first time in history China exerted control over the Tibetan economy and Tibetan resources. While economic ties have existed between Tibet and China for hundreds of years these had previously been limited to trade and were mutually advantageous. State ownership of Tibet's richest natural resources and imposition of policy from the central government have precluded any semblance of economic autonomy for Tibet. Tibetans are generally excluded from decision-making over the exploitation of the resources and land that they depend upon for their livelihoods and, in practice, there has been only minimal trickle-down of benefits to Tibetan nomads and farmers who make up the majority of Tibet's population. Mining has most often become relevant to Tibetans when they have had to live with the long-term social and environmental costs.

Many of the issues discussed in '*Mining Tibet*' parallel similar problems in other countries where mining has prompted debate over the rights of indigenous communities in relation to mineral exploitation. With few exceptions, local communities tend to lose out to the interests of national authorities and mining companies, and face economic, social and cultural marginalisation as a result of mining development. Issues of control over land and resources are particularly sensitive when local people belong to a different ethnic group from the dominant nationality of the governing state. These sensitivities can become heightened when a significant portion of the impacted ethnic group does not identify itself with that state. These concerns are becoming more acute for Tibetans and other 'minority nationalities' in China as the authorities drive forward 'leap-over style (*kua yue shi*)' development.

Mining and its impact on Tibet are also becoming increasingly relevant to the international community as China solicits foreign capital and technological investment, seen as essential to implementing development strategies in the western regions.

Attracting foreign investment in mineral exploration and extraction in Tibetan areas remains a challenge due to the remoteness of the territory, poor infrastructure, bureaucracy, an inadequate regulatory, legislative and judicial regime, and international as well as domestic political sensitivity. Nonetheless, economic reform in China continues to move at a fast pace, interest is growing and foreign involvement in the industry's development in Tibetan areas is increasing.

The environmental impact of mining in Tibet is an issue of global consequence. The Tibetan plateau is one of the world's most significant high plateau upstream environments, and the impact of development in Tibet extends to Asia and beyond. The Yangtze flood disaster in 1998, linked to soil erosion and deforestation on the Tibetan plateau, was a powerful wake-up call to the Chinese leadership and many Chinese intellectuals and environmentalists are now expressing concern about 'fast-track' development policies based on exploitation of the fragile upstream environment. The impact of mining – both legal and illegal, state-funded and private – on Tibet's ecosystem provides a vivid illustration of the conflict between economic development and environmental protection, a pressing concern in modern China and throughout the world.

It is difficult to get a full picture of the mining industry in Tibet. On the one hand there is no free flow of information in China, while on the other hand official reports on development achievements or potential tend to be exaggerated or skewed to create an impression of success for state policies and initiatives. While it appears from official economic data that large state subsidies are essential for construction, such data does not place a value on past or potential future gains from state exploitation of mineral and other natural resources in Tibetan areas. The commonly held perceptions of Tibet as an unspoilt, ecologically pristine land and also as a depository of minerals ripe for exploitation both mask the reality of the development that has taken place over the last 50 years. While the remoteness of the land, harsh climate and altitude have prevented extraction of many minerals across much of Tibet, other areas have already been extensively exploited.

The purpose of this book is to provide a comprehensive study of the history, development and future prospects of the mining industry in Tibet within the wider context of China's reform and development, and to examine the impact that this has had on Tibetans and their environment. The wide variety of primary and secondary material on which this study has been based includes: interviews with Tibetans carried out in Nepal and India; consultation with various specialists; Chinese government policy documents, including a two volume internal (*neibu*) document outlining the TAR's plans for resource use 1996-2020; PRC laws and regulations; official

Chinese statistical data; Chinese newspaper reports; investment project summaries; Chinese academic journals; and a variety of English-language source materials including articles, conference papers and books, data and reports from the international mining industry and government and international agencies, NGO project reports, international law and standards; as well as extensive use of Chinese and English language websites and search engines.

Note on Geography

There is still considerable confusion in the West as to what constitutes 'Tibet'. Tibet was traditionally comprised of three main areas: Amdo (north-eastern Tibet), Kham (eastern Tibet) and U-Tsang (central and western Tibet). The Tibet Autonomous Region (Ch: Xizang Xizhiqu) was set up by the Chinese government in 1965 and covers the area of Tibet west of the Yangtse River, including part of Kham, although it is often referred to now as 'Central Tibet' in English. The rest of Amdo and Kham have been incorporated into Chinese provinces, and where Tibetan communities were said to have 'compact inhabitancy' in these provinces they were designated Tibetan autonomous prefectures and counties. As result most of Qinghai and parts of Gansu, Sichuan and Yunnan provinces are acknowledged by the Chinese authorities to be 'Tibetan'. The term 'Tibet' in this report is used to refer to all these Tibetan areas currently under the jurisdiction of the PRC (see map on inside back cover).

Of necessity, this report reflects the political geography as defined by the Chinese state. TIN recognises that many geographic labels in modern usage do not adequately reflect traditional Tibetan views on regional nomenclature or on political administration. Where possible, Tibetan names are used for place names within administrative entities designated 'Tibetan autonomous' by China. On the first use, the Chinese equivalent will be provided in brackets. A list of the place names (county-level and above) mentioned in this book is provided in the back pages in Tibetan script and Chinese characters.

Summary Points

- The Chinese authorities started surveying and mining Tibet in the 1950s. The mining industry expanded considerably due to the economic reforms of the 1980s and 1990s, but generally remained small-scale and inefficient. As China faces growing shortages in the domestic supply of raw materials the urgency to accelerate large-scale exploitation of Tibet's minerals has increased and as infrastructure develops key mineral commodities are becoming increasingly accessible.
- Despite the state's ambitious plans there are many obstacles to development of the mining industry including; Tibet's remoteness, high altitude and harsh climate; inefficiency and corruption; unreliable geological data; and an inadequate legislative, regulatory and judicial regime. Attracting investment, particularly from larger foreign mining companies, remains a challenge for authorities.
- Poor governance and control over mining have exacerbated its environmental impact. The central authorities have lost state assets and revenue through extensive rather than intensive mining practices and illegal mining and trading. The interests and concerns of officials and local people have been subordinated to the personal interests of officials at higher levels.
- Key concerns for Tibetans are expropriation of pasture and arable land, in-migration, and damage to the local environment that affects livelihoods, health and also transgresses religious beliefs and cultural norms. The response of authorities to complaints, petitions and direct action ranges from compensation to criminal prosecution. Mining development has added to general dissatisfaction with Chinese rule and sometimes been an element in political protest.
- The mining industry in Tibet is dominated by ethnic Chinese, particularly in skilled and managerial sectors. In basic jobs Tibetans face increasing competition from Chinese workers. Although the authorities state that they support vocational training for Tibetans, they are encouraging the import of 'qualified' personnel from eastern China to accelerate development. As a result mining is exacerbating existing economic and social disparities between Chinese and Tibetans in Tibet.
- Conditions in Tibet's mines appear to be consistent with the notoriously low health and safety standards throughout China. Over the past 50 years local communities and prisoners have been an important source of manual labour within mining enterprises in Tibetan areas and for preparing supporting infrastructure.
- Mining has already had a serious environmental impact in some Tibetan areas, notably land degradation, pollution and harm to livestock and wildlife bio-diversity. Growing awareness of this amongst China's leaders is offset by conflicts of interest, both political and economic, at central and local levels. Despite major advances in developing a legislative framework to regulate environmental protection, implementation remains very limited.
- Mineral resource-based development appears to be increasing rather than decreasing Tibet's reliance on subsidies, and the gap between rich and poor is growing. The majority of Tibetans (who are mostly nomads and farmers) have seen few benefits and Tibetan communities living with the environmental and social impacts of mining face increasing marginalisation.

1. Policy and objectives

The exploitation of natural resources is at the heart of China's plans for the economic development of Tibetan areas and the other western regions and provinces under the PRC's administration. The developmental model chosen for these areas serves the national interest in meeting domestic demand for raw materials. There are obvious disadvantages to large-scale extraction on the Tibetan plateau, including remoteness and poor infrastructure. The authorities are also becoming increasingly aware of the importance of protecting Tibet's fragile environment.[1] Despite these disadvantages, the strength of factors driving the desire to open up Tibet's resources have led the authorities to push forward with plans to accelerate the development of the mining industry in Tibet. As mineral reserves in the eastern and central parts of the country are being depleted, the urgency to speed up development in Tibetan areas has increased, and China is aiming to move towards organised large-scale exploitation.

The first section of this chapter outlines the economic policy of the Chinese authorities, which remains committed to accessing the minerals of the Tibetan plateau. It looks at the current stage of mining development in Tibet and the growing possibilities for the near future. This section also provides a brief overview of the historical development of the mining industry in Tibet, from the national construction efforts of the 1950s and 1960s and through the economic liberalisation of the 1980s and 1990s.

The second section consists of three case studies that explore in greater detail the history and future of the mining industry. The first looks at the development of mining in the Tsaidam basin in Tsonub Mongolian and Tibetan Autonomous Prefecture, Qinghai province, a vast region that prior to the 1950s was sparsely populated by Mongolian and Tibetan nomads but is now the most important mining area in Tibet. The second case study is of gold mining, the most widely exploited mineral across Tibetan areas. It establishes the continued importance of gold as a commodity in China, before charting the history and development of gold mining in Tibet and assessing its extent and importance in relation to China's gold production and future demand. The third case study looks at copper, a mineral commodity in very short supply in China, and of which Tibet has the second largest reserves in the country. The aim is to provide a detailed analysis of a key commodity, previously unfeasible to develop, that is now being opened up to exploitation.

1 The obstacles to the development of the mining industry are looked at in detail in Chapter 2.

Policy and historical development

Policy aims

The central authorities remain firmly committed to developing the mineral resources of Tibet to support national development and industrialisation, despite the many challenges they face in achieving this. China currently faces a shortage in the supply of several mineral resources vital to the economic growth and industrialisation of the whole nation, including petroleum resources, gold, copper, high-grade iron ore, chromite, cobalt, platinum and manganese. One of the key aims of the current drive to develop the western regions of China is to access deposits of these minerals that have previously remained out of reach.

China's 10th Five-Year Plan outlines the key state aims and projects over the five years (2001-2005). According to the plan, the need to intensify exploitation of natural resources to meet domestic demand is of particular urgency, as is the need to 'forge ahead aggressively' with the development of the western regions.[2] The 'priority projects' that China is concentrating on during this period include the construction of the Golmud-Lhasa railway, which will link central Tibet to Qinghai and beyond to the Chinese interior. Other planned reforms during the Tenth Five-Year Plan allow for increased rural to urban population migration and intra-regional population movements. Large-scale infrastructure construction, greater mobility of labour and urbanisation are vital for China's plans to exploit the western regions' resources. Regional and provincial development plans are constructed accordingly, based around this central model.

The central government's policy to accelerate development in Tibetan areas, formulated a decade ago, focuses on GDP growth and the exploitation of resources for industrialisation and development.[3] One of the main aims has been to improve the infrastructure of these areas, to open them up to development and exploitation and achieve Tibet's closer economic, political and social integration into the Chinese state. The drive to develop China's western provinces and regions, officially launched in June 1999, marked an intensification of state efforts to fulfil these aims and an attempt to attract domestic and foreign investment in support. Preferential policies have subsequently been released to encourage investment in China's western regions, including in resource exploitation, and to subsidise infrastructure construction.[4]

2 China's Tenth Five-Year Plan Outline, Office of the Premier, State Council of the PRC, Beijing, in Chinese 5 March 2001, translated by BBC Monitoring.

3 & 4 See bottom of next page

However, until now the mining industry has largely remained small-scale andinefficient in Tibetan areas, as it generally has throughout China. Relatively few of the mineral resources of Tibetan and the other western regions of China have been exploited as a result of the area's remoteness and poor infrastructure. The central government has lost considerable revenue and state resources as a result of extensive rather than intensive mining practices,[5] local protectionism and illegal mining,[6] and is now attempting to reassert control over mining to maximise the potential of state resources.

Similar to industry throughout China, there are increasing pressures on the minerals industry to become more efficient and more competitive, particularly as China becomes further integrated into the global economy and faces competition in an international minerals industry that is currently marked by falling prices. As a result the central authorities aim to modernise and mechanise the mining industry in Tibet and, similar to industrial restructuring throughout China, to develop larger conglomerates and close down smaller unprofitable enterprises.

The Chinese authorities are now calling for a 'leap-over style [Ch: *kua yue shi*]' development in Tibetan areas, as opposed to simply 'fast-track' or 'accelerated' development. Improvements in infrastructure are increasing the feasibility of large-scale resource exploitation in at least some areas of Tibet and have facilitated geological exploration.[7] Sustained development over the next ten or so years is likely to significantly alter the economics of ore extraction on the Tibetan plateau, particularly once the Golmud-Lhasa railway is opened. This, combined with the central government's desire to control and limit small-scale mining, and the push to meet domestic demand for raw materials, reflects the aim of the authorities to move towards organised large-scale exploitation of Tibet's resources.

3 The macro model for current economic development throughout Tibetan areas reflects the policy officially set for the TAR at the Third Forum on Tibet Work, held in Beijing in 1994. The forum provided a central mandate for the fast-track economic reforms that had been initiated in the TAR in 1992. The priorities of the authorities to push forward large-scale infrastructure construction and key state projects were reflected in the '62 projects' that were announced at the Third Forum and funded by the central government and other Chinese provinces and municipalities. The largest proportion of investment went into power construction, mainly hydropower – in total 14 of the 62 projects. A further nine were also infrastructure construction projects (roads, pipelines and telecommunications). Four were mines: the Xiangka mountain chromite mine in Chusum (Ch: Qusong) county, Lhoka (Ch: Shannan) prefecture; the Bengnazangbu gold mine in Shantsa (Ch: Shenza) county, Nagchu (Ch: Naqu) prefecture; a Szaibelyite (magnesium borate hydroxide) mine at Chagzam salt lake; and the Machala coal mine in Riwoche (Ch: Leiwuji) county, Chamdo (Ch: Changdu) prefecture. In comparison, only five of the 62 projects were agricultural, although six of nine industrial projects were concerned with agricultural and animal products processing. (For a full list of the projects see TIN News Review: Reports from Tibet, 1998, London: TIN, 1999, pp. 90-91)

4 See Chapter 2

5 'High grading', see chapter 2, pp. 57-58

6 See chapter 2, pp. 54-63 and chapter 5, pp.162-170

7 See Chapter 2, pp. 46-49

Historical development of the mining industry in Tibet

> *"After liberation, following the development of the forces of production and the needs of economic construction, requirements for mineral resources increased unceasingly".* TAR National Land Specialist Plan (1996-2020).[8]

Prior to the 1950s, Tibetan lands and resources were owned by the central Tibetan Government (Ganden Phodang), or the local rulers of those areas outside Lhasa's control. Exploitation of most mineral resources was minimal and there was no mechanised industrial-scale mining in Tibet, although Tibetans are known to have engaged in gold and salt mining for hundreds of years and Chinese prospectors also entered the eastern fringes of Tibet to mine gold and silver.

The incorporation of Tibetan areas into the People's Republic of China (PRC) brought with it the Chinese state's assertion of ownership of all mineral resources within its territory, and mining became a state industry. During the 1950s and 1960s, government and military exploration teams were sent into Tibetan areas and the exploitation of Tibet's resources was worked into national development plans. The main area of Tibet where the basis for a mining industry became most firmly established during this period was the Tsaidam Basin, which over the last fifty years has developed into the most important area for mineral resource exploitation on the Tibetan plateau (see case study below).

The initial focus of the authorities was on the development of energy resources, coal and oil, to support early industrialisation and China's national construction efforts. According to the Tibet Autonomous Region (TAR) National Land Specialist Plan (1996-2020), an internal (*neibu*) document setting out official plans for the development of regional resources including minerals, the first coal mining district in central Tibet was established in Tumengela in Amdo (Ch: Anduo) county, Nagchu (Ch: Naqu) prefecture, in the early 1960s.[9] Smaller coal mines were also established in Chamdo (Ch: Changdu), Shigatse (Ch: Rigaze), Ngari (Ch: Ali) and Lhasa.[10] Boron deposits were also developed in central Tibet in the late 1950s and early 1960s, including the notorious Jangthang Borax Mine – a labour camp where many Tibetans arrested following the 1959 Tibetan Uprising were sent for 'reform through labour'.

8 '*Xizang Zizhiqu Guotu Zhuanti Guihua* (1996-2020)', a two volume internal (*neibu*) document formulated by the TAR National Land Plan Formulation Committee (guotu guihua bianzhi weiyuanhui) and the TAR Planned Economy Committee (*jihua jingji weiyuan hui*). Although undated, it is clear that the document was written after the Third Tibet Work Forum (1994), and is believed to have been issued 1994-1995. Referred to below as the 'TAR Specialist Plan', p. 464.

9 'Central Tibet' is used here to refer to the area that was officially established as the Tibet Autonomous Region (TAR) in 1965.

10 TAR Specialist Plan, op cit, p.464.

Ferrochrome resources were mined at the Dongfeng Iron Mine in Chusum (Ch: Qusong) county, Lhoka (Ch: Shannan) prefecture, established in 1967, and later to become the Norbusa Chromite Mine, now the largest chromite producer in China. Gold was also mined in Tibet during the 1950s and 1960s, starting as early as 1954 in Dege county, Kardze (Ch: Ganzi) Tibetan Autonomous Prefecture (TAP), Sichuan province, when the Sichuan-Tibet highway (National Highway 317) was opened.

However, state exploration and exploitation of Tibet's mineral resources remained constrained by the remoteness and poor infrastructure (power, telecommunications, roads) of the Tibetan plateau. For example, the authorities in Dartsedo (Ch: Kangding) county, Kardze TAP, attempted to establish copper mines in 1958 and 1959, but on both occasions had to close the mines soon after becoming operational as a result of lack of fuel for smelting and poor transportation links.[11]

The far-reaching economic reforms that have taken place throughout China following the death of Mao Zedong in 1976 have had a significant impact on the extent of the mining industry in Tibet. State efforts to open up the vast sparsely populated Tibetan plateau during the 1980s and particularly the 1990s, have provided a basis for the development of Tibet's mineral resources. Infrastructure construction and immigration have already changed the face of many Tibetan areas, particularly those situated on the fringes of Tibet and along the Xining-Golmud railway, and it is these areas that have been most heavily exploited for minerals.

From the late 1970s, the state renewed its efforts to survey the Tibetan plateau in order to discover and map new resources and verify and develop existing reserves. China's national and regional five-year plans have included key mining projects in Tibet marked for state investment. For example the Xitieshan lead and zinc mine in the Tsaidam Basin (Sixth Five-Year Plan), the Norbusa Chromite Mine in Lhoka prefecture, TAR (Eighth Five-Year Plan) and the Yulong copper mine in Chamdo prefecture, TAR (Tenth Five Year Plan).

However, since the 1980s, there has also been a massive increase in the types of unit engaging in mining. State-owned enterprises have remained responsible for larger-scale mining enterprises, but government departments and organisations at every level, township collectives, the military and police have all been encouraged to exploit small-scale deposits. During the 1990s in particular, mining has been promoted as a way for local governments to demonstrate their entrepreneurial skills and generate income.

11 Kardze Annals Editorial Committee (Director Liu Yingjie), 'Kardze Annals *(Ganzi Zhouzhi)*', Sichuan People's Publishing House, Chengdu, 1997, p.1181.

During the 1990s, small-scale private mining enterprises were encouraged in Tibet and provincial/regional, prefectural and county governments were urged to widen investment sources. By the end of the 1990s, the Mineral Resources Law had been amended to allow for private domestic and foreign investment and authorities at every level were seeking investment for mining projects both internally from domestic mining companies, but also from foreign companies. 'Counterpart aid' programmes were also established with other Chinese provinces, required to provide technical assistance and financial investment.

The most widespread development of mining throughout Tibet areas appears to have been the massive increase in small-scale gold production during this period. Gold is relatively cheap and easy to mine, and as a result became one of the few minerals that local governments and collectives could feasibly exploit. During the 1990s local governments also started leasing land out to private mining enterprises. An unofficial gold industry also blossomed during the 1980s and 1990s, as tens of thousands of prospectors entered Tibetan areas to mine gold, most of them poor rural Chinese or Hui Muslims from neighbouring provinces, although Tibetans have also engaged in artisanal mining.

By the end of the 1990s mineral extraction and processing enterprises throughout Tibet included gold, copper, coal, chromite, asbestos, aluminium, boron, potassium, salt, lead and zinc. In north-eastern Qinghai province, Pari Tibetan Autonomous County (TAC) in Gansu and the Tsaidam Basin, minerals were being heavily exploited. However, the development of mining in most other areas of Tibet remained constrained and, although increasing in extent over the 1980s and 1990s, production has remained small-scale with the exception of a few key enterprises. Throughout Tibetan areas mining enterprises have generally been inefficient and unmodernised – similar to most mining enterprises throughout China – and foreign investment has remained minimal.[12]

12 See section on investment, including foreign investment, chapter 2, pp. 65-75

Gold-panning dish © TIN

Asbestos processing plant and products factory, Chilen (Ch: Qilian) county, Tsojang (Ch: Haibei) TAP, Qinghai
© TOTAR, 1997

Case studies

The Tsaidam Basin

The Tsaidam Basin is located in Tsonub (Ch: Haixi) Mongolian and Tibetan Autonomous Prefecture (M&TAP), which occupies the north-west area of Qinghai province. Until the 1950s, the Tsaidam Basin had a sparse population of predominantly Tibetan and Mongolian nomads. However, it is considered to have enormous potential for exploitation and over the past 50 years it has developed a basic minerals industry as a supplier of raw materials from its salt lake resources and its rich deposits of petroleum, non-ferrous metals and construction materials, including asbestos. By 2001, the Tsaidam Basin had 32 urban centres, most of them developed either directly or indirectly in relation to the mining industry.[13] Tibetans and Mongolians have become a small minority of the population, which according to official Chinese data is now dominated by Han Chinese and, over the last ten years, a rapidly growing Hui Muslim population.[14]

As early as 1953 the Chinese government had dispatched geological survey teams to explore the area, and as a result of their findings the central authorities decided to prioritise industrial development over the development of pastoral areas.[15] Oil prospecting began in 1955, and extensive oil reserves were discovered at Lenghu in 1959. These became one of the principal petroleum resources within the PRC, and of great importance in supplying the development needs of the new bases for settlement and exploitation in Amdo and Central Tibetan regions.[16]

In addition to petroleum, the Tsaidam Basin is most famous for its salt lake resources. The area has 33 salt lakes, including Cha'erhan, situated north of Golmud which has been mined since 1958. According to the Qinghai Salt Industry Corporation, it is now believed to have the largest soluble potassium-magnesium deposit in China and one of the largest in the world. Other resources initially developed in the 1950s include the lead and zinc mine at Xitieshan in Dachaidan, which has been in operation since 1957. Coal mines were also developed to provide power for agricultural labour camps (Ch: *laogai*), new settlements and industrialisation.[17]

13 Cun Wanjuan and Zhang Zhimin (Qinghai Provincial Hydropower Department), 'Water resources and the construction of small towns in Tsaidam', Tsaidam Development Research (*Chaidamu Kaifa Yanjiu*), Vol.1, 2001

14 See Chapter 3, note 27

15 June Teufel Dreyer, 'Ch'inghai', in Edwin A.Winckler (ed.), A Provincial Handbook of China, Stanford University Press, c.1975, cited in Steven D. Marshall and Susette Cooke, 'Tibet Outside the TAR', CD-ROM published by the authors, 1997, p.1861

16 Marshall and Cooke, 1997, op cit, p.1862

17 James D. Seymour and Richard Anderson, 'New Ghosts, Old Ghosts' , New York and London: M.E.Sharpe, 1998, p.157

The development of a basic mining industry and urban infrastructure in the Tsaidam Basin has required colonisation. In the 1950s and 1960s development of both agriculture and industry was largely dependent on the forced resettlement of population from inland China, the 'reform through labour' of prisoners and intellectuals who were put to work in the sprawling labour camp (*laogai*) network and the military.[18] Construction of settlements in the mining areas of Dachaidan and Mangya, using the labour of prisoners and forced immigrants, began in 1956.[19] The Mangya asbestos mine, which is reported to have the largest asbestos reserves in China, was opened in 1958. Steve Marshall and Susette Cooke, who have carried out extensive fieldwork in and research on Tibetan areas outside the TAR commented that:

> *"More than any other Tibetan area, Tsonub has been treated by the Chinese as a giant laogai facility, occupied and controlled for the prime purpose of extracting rich mineral resources."*[20]

The economic reforms of the 1980s and 1990s have accelerated the development of mining areas and the colonisation of Tsonub M&TAP and the Tsaidam Basin. One of the key developments that has facilitated the development of Tsonub's resources over the past 20 years was the opening of the Xining-Golmud railway in 1984. This link through to the provincial capital of Xining and beyond to the Chinese interior has significantly altered the economics of mineral exploitation in the area. Policies to encourage voluntary migration have provided a flow of migrant labour to the mining and construction industries that were originally founded on forced labour.

Although the Tsaidam Basin has the most concentrated and developed mining industry of any Tibetan area, exploitation still remains relatively basic and inefficient.[21] However, development of Tsonub and the minerals of the Tsaidam Basin are being pushed forward under the Western Development Campaign and there is a desire to attract domestic and foreign investment into the region. There appears to be an increasing emphasis on the establishment of processing and manufacturing centres based around Golmud, although many of the raw materials still appear to be transported out to Xining. One large-scale construction project, the Sebei-Xining-Lanzhou gas pipeline, was operational by the end of 2001. Italian Petroleum Company Agip became the first foreign mining or petroleum company to operate in Tibet when it signed a contract in April 2002 with China National Petroleum Corporation to become operator of the gas-fields in the Sebei Block of the Tsaidam Basin.[22]

18 See Chapter 3, pp. 87-88
19 Marshall and Cooke, 1997, op cit, p. 1861
20 Marshall and Cooke, 1997, op cit, p.1871
21 For example see Cun Wanjuan and Zhang Zhimin (op cit) or Song Xinyu and Yao Jianhua, 'Resources Exploitation and Conservation of Qaidam Basin under the Principle of Sustainable Development', a paper presented to the International Symposium on the Qinghai Tibet Plateau, Xining, 24 July 1998
22 'Agip drilling on its northwest China Sebei Block', Oil and Gas International, 23 April 2002

Gold mining

Demand for gold in China

Even before the Chinese Communist Party (CCP) came to power in 1949, they had established a state monopoly on gold and silver in areas under their control because of the direct impact these metals have on monetary stability. Until relatively recently, the PRC was reliant on gold as foreign exchange and gold accounted for a high proportion of the country's financial reserves. Now gold only represents three per cent of China's financial reserves and the state is moving towards relinquishing its controls on gold and silver as part of the wider market reforms being undertaken in China. The international gold price has been in decline for the past two decades and demand in Asia fell at the end of the 1990s due to the economic crisis in the region.

However, gold remains an important commodity in China, valuable not only for its financial function, but also as a raw material for the development of a high-tech industry (gold has high conductivity). There are also hopes for a significant increase in demand for gold jewellery amongst domestic consumers, spurred on by the World Gold Council which is reported to be actively promoting gold jewellery sales via television soaps, and gold bullion remains a safety net against Renminbi devaluation.[23]

The gold industry in China has faced serious challenges to its long-term growth from 'reckless mining', backward technology and the resultant waste and low efficiency, not to mention a huge loss of revenue to the black market.[24] The precise amount of gold mined throughout China is unknown even to the Chinese authorities. According to official statistics, China's gold output has increased dramatically over the past ten years, from an estimated 100 tonnes in 1990[25] to over 180 tonnes in 2001.[26] However, these official figures do not necessarily reflect the real amount of gold being mined within China. Official pricing policies that have alternated between keeping the price of gold above or below the international market level have resulted in significant quantities of gold being smuggled into and out of the country and mines are also believed to have been stocking gold in anticipation of domestic price rises.[27]

23 Executive summary of John Adams, 'Deregulation of the Gold Market in China', 2001, on www.chinaonline.com and John Adams, 'A New Player?', FEER, 20 December 2001

24 See Chapter 2, pp. 59-63

25 Pui-Kwan Tse, 'The Mineral Industry of China, 1994', USGS, 1995.

26 China GoldNet, 23 January 2002, www.gold.org.cn. South Africa is the world's largest gold producer although output has been falling since 1990. According to the South African Chamber of Mines, South Africa's gold output was 423.3 tonnes in 2000, the lowest production level since 1954 ('South Africa's Gold Output Falls', Gold Information Network, 22 February 2001).

27 John Adams, op cit.

However, what is certain is that China is still not meeting domestic demand for gold, estimated to be over 200 tonnes a year.[28] Moreover, gold production could actually fall over the next few years as a result of current reforms, which include the closure of small mines. According to John Adams, director of China Financial Services and author of a report on China's gold market, China's reported proven gold reserves will only give supplies for two to three years, while new capacity is only adding 10-15 tonnes per year. *"China"*, he concludes, *"may therefore face stagnating or declining levels of domestic supply"*.[29]

In September 2000, Wang Dexue, director of the Gold Administration Bureau of the State Economic and Trade Commission (under the State Council), gave a speech on the reform and opening up of China's gold industry. He said that the most serious problem in the development of the industry is the shortage of gold resources and that the solution lies in developing China's western regions. Most gold enterprises are in east and central China and following 'ten years high speed development' increase in production is slowing down. He went on to say that: "*If China's gold industry wants to develop continuously in the 21st century new areas must be opened up, that is to set off a rush for gold development in broad western China*".[30]

Gold mining in Tibet

Gold has been mined in Tibetan areas for hundreds of years. The Indian pundits, sent by the British administration in India to make secret surveying missions to Tibet in the 19th century, brought back reports of the gold fields being mined in western Tibet,[31] while accounts in Chinese sources describe gold mining in eastern Tibet[32] going back two to three hundred years. However, since the 1950s exploitation has been more extensive. In particular, mass small-scale and artisanal mining during 1980s and 1990s, and the mechanisation of some operations during the 1990s, has dramatically altered the reach and effects of small-scale gold mining.

During the past fifty years, gold has been the mineral most widely exploited throughout Tibetan areas, although generally on a small-scale. Up to the present time, most gold mining in Tibetan areas has been of the alluvial deposits (i.e. gold found in sediment that has been deposited by flowing water) that have collected in riverbeds or under the pastureland. Alluvial gold is relatively easy and cheap to mine and where prospectors are successful it can bring a quick turn around of profit.

28 Liu Shanen, 'The progress in the reform for gold marketing in China', Beijing Research Centre for the Development of [the] Gold Economy, September 2000, speech published on China GoldNet. According to Gold Field Mineral Service, demand reached 184 tonnes in 2000, cited in John Adams, 2001 op cit.
29 John Adams, op cit.
30 Wang Dexue, 'Three Issues Facing China's Gold Industry in New Century' September 2000, on China GoldNet www.gold.org.
31 Peter Hopkirk, 'Trespassers on the Roof of the World: The Secret Exploration of Tibet', Kodansha, 1995.
32 Traditional area of Kham. Now mostly incorporated into northern Yunnan, western Sichuan and eastern TAR.

As governments were encouraged to develop local mining through collective and government enterprises from the 1980s, gold was the principal mineral resource feasible to exploit. For example, Kardze TAP in Sichuan province launched a gold production drive at the beginning of the 1980s. Various military, prefectural, provincial and state geological surveying teams had entered Kardze from 1955 onwards and prospecting for gold started soon after the PRC took control over the area. However, efforts were intensified as a result of reforms in the 1980s, involving the opening of prefecture and county-level government mines and village and township enterprises. By the beginning of the 1990s, relatively large gold mines (although still small-scale) were concentrated in counties including Dartsedo, Payul (Ch: Baiyu), Serthar (Ch: Seda), Kardze and Dabpa (Ch: Daocheng).[33] During the 1990s, there was also a growth in private gold mining in Kardze, with local authorities leasing land to entrepreneurs on short-term contracts. Gold has also been mined by prisoners detained in the prefectural prison/laogai network. During the Ninth Five-Year Plan period (1996-2000) there were plans to build more gold production bases in Dartsedo and Serthar.[34]

During the 1980s, as economic liberalisation allowed greater freedom of movement for China's rural poor, an unofficial rush for gold also started in Tibetan areas. Tens of thousands of Chinese and Hui Muslim peasants and labourers entered Tibetan areas to set up makeshift camps by riverbeds or on pastureland, mining gold and hunting as a form of poverty alleviation. The largest concentrations of these prospectors have been in Nagchu prefecture in the TAR, Kardze TAP in Sichuan and Yushu TAP in Qinghai and some of the areas they have mined are of particular ecological significance.[35]

By the end of the 1990s gold mining had become an important source of local government income in some Tibetan areas. As well as establishing their own small-scale gold mining enterprises, as county or township enterprises, local governments also obtained income from quasi-legal and illegal gold mining, profiting from selling mining permits, buying gold from prospectors and unofficial trading. Although the authorities continued to encourage localised exploitation for economic development during the 1990s, central concern grew over the extent to which gold mining was out of central control in Tibet and throughout China.[36]

Despite the apparently high level of activity in Tibetan areas, the gold output of the TAR and Qinghai, Sichuan and Yunnan provinces does not appear to have been that significant in national terms.[37] Mining enterprises have remained largely small-scale, while much of the gold mining that has taken place throughout Tibetan areas appears to have been localised, out of state control and either illegal or quasi-legal.

33 Kardze Annals, op cit.

34 Zhongguo Xinwen She, 27 August 1997; published in English by BBC Monitoring, SWB, 1 September 1997.

35 See Chapter 5

36 See Chapter 2, pp.59-63 and Chapter 5, pp. 171-179

Medium-large scale mines

The largest gold producer in Qinghai province is reported to be the Tanjianshan bedrock gold deposit, located in Dachaidan, Tsonub M&TAP. The deposit, which has estimated reserves of 52 tonnes,[38] was discovered in 1989 and is now being developed by the Dachaidan administrative committee (county-level) and Qinghai's No.1 Geology Exploration Unit who have jointly formed the Dachaidan Jinlong Mining Development Company.[39] The mine appears to have been operational on a basic level since the early 1990s, and had produced over one tonne of gold by 1999. However, once fully operational, the deposit is expected to produce as much as 2.8 tonnes of gold per year.[40] It attracted serious interest from Australian company Sino Mining International who invested US$2.1m in prospecting. However, following completion of their first phase exploratory drilling program, the company pulled out in February 2001, stating that the project would not be sufficiently profitable to warrant further exploration or development at that time.[41] Sino Mining has since formed a joint-venture with a Chinese company to explore the Jinkang gold deposit in Kakhog (Ch: Hongyuan) county, Ngaba (Ch: Aba) Qiang and Tibetan Autonomous Prefecture (Q&TAP), Sichuan.[42]

According to data provided in the TAR National Land Specialist Plan (1996-2020) there is only one large alluvial gold mine in the TAR, the Bengnazangbu mine in Shantsa (Ch: Shenzha) county, Nagchu (Ch: Naqu) prefecture.[43] The mine was one of the 62 'Aid-Tibet' projects announced at the Third Tibet Work Forum in 1994, and went into operation in 1997, with a designed capacity to produce 386 kg (about a third of a tonne) of gold per year. The mining area is reported to have proven (Ch: *chuliang*) reserves of 10 tonnes of alluvial gold. In 1996, Tibet Daily announced that another two deposits would be developed in the area before 2005, each with annual gold production of over 200 kilos and 20,000 tonnes of associated antimony ore.[44] According to an exploration geologist who has examined the data in the Specialist Plan, the TAR's other alluvial deposits are evidently small and probably uneconomic to mine by any organisation or company.

37 According to official gold mine production figures, China produced 175 tonnes of gold in 2000. Of this, 115 tonnes was produced by gold mines (the rest was produced by refineries). Shandong's output accounted for 23 tonnes of the total 115 tonnes, more than twice the next largest producer, Henan, with 11 tonnes. The western regions and provinces that come under Western Development Campaign plans produced a total of 36 tonnes (or 31 per cent of the national output). Those incorporating Tibetan areas accounted for 13 tonnes (11% of the national total), of which the Tibet Autonomous Region (TAR) produced one tonne, Qinghai three tonnes, while Gansu produced five tonnes and Sichuan and Yunnan both produced two tonnes. A more recent report indicates that gold production levels have fallen in these provinces/region in 2001, with the exception of Yunnan. 'A Profile of China's Gold Industry', Asia Pulse Analysts, Update Nov 2001, on China GoldNet china.gold.org.cn.

38 'China Business Handbook 2002', China Economic Review, 2002

39 Dachaidan Jinlong Mining Development Company introduction.

40 Sino Mining Limited, project information, 2000, on the SMI website www.sinomining.com.au

41 'Sino Mining has no current plans to develop mine asset in Tibet', Australia Tibet Committee, 5 February 2001

42 See Chapter 2, p.74

43 & 44 see next page

The largest known gold mine operating in Kanlho (Ch: Gannan) TAP, Gansu is the Machu (Ch: Maqu) county gold mine. In operation since the early 1990s, the target for the mine was to produce 500kg (half a tonne) by 1997. Reports from the area suggest that further deposits are being explored and developed in the area. The aggregate output of the seven operational high-yielding mines Kardze TAP and Ngaba Q&TAP already contributes significantly to Sichuan's provincial gold production, accounting for nearly 50 per cent of Sichuan's total output in 1997.[45] However, given that Sichuan's total gold output was about two tonnes in 2000 and 2001, these mines can not be considered to be major contributors to national gold output.

Future development

In terms of future development of the gold industry in Tibetan areas, the main priority of the higher authorities appears to be to consolidate medium to large-scale productive mines and clamp down on small-scale and artisanal gold mining. This reflects central aims to reform the gold industry throughout China. The first step towards achieving this in Qinghai came in February 2002, when the provincial authorities issued a ban on all alluvial gold mining. The only exceptions to this were two mines in Pema (Ch: Banma) county in Golog (Ch: Guoluo) TAP and the Zhaduo gold mine in Tridu (Ch: Changduo) county, Yushu TAP.[46] If planned reforms are pushed through there is likely to be a drop in gold production in Tibetan areas in the short-term, until better organised and larger-scale exploitation can be achieved.

The authorities are also looking to discover and develop new deposits to meet the shortage of gold reserves in China. Although alluvial gold resources have been mined extensively in many areas, Tibet's bedrock gold resources have generally not been exploited. They are potentially more profitable than alluvial resources, but are more difficult and costly to extract and require larger-scale operations.

Based on available information, the main commercially prospective areas within Tibet appear to be Ngaba Q&TAP in Sichuan and Kanlho TAP in Gansu. A recent project to comprehensively document China's gold resources, has mapped the existence of 33 bedrock gold deposits in the 'Western Qinling Belt', 15 of which are located in Ngaba and Kanlho, while one is in Kardze TAP.[47] The project was a collaborative venture between the Chinese government, several foreign mining companies and the US Geological Survey, as well as Western and Chinese academics, and its results were published in two papers that appeared in the journal 'Mineralium Deposita'.[48]

43 (see previous page) TAR Specialist Plan, op cit.
44 (see previous page) Tibet Daily, 13 December 1996. The same year, TAR branch of the Agricultural Bank of China provided financing for technological upgrading (Xinhua 4 July 1996).
45 Zhongguo Xinwen She, 27 August 1997; op cit.
46 'Qinghai province bans gold digging to protect the environment', Qinghai Daily, 9 February 2002
47 & 48 See next page

Three of the deposits, located in Luchu (Ch: Luqu), Machu and Zungchu (Ch: Songpan) counties are of particular interest to prospectors. One of them, the Dongbeizhai deposit in Zungchu is the largest known gold resource in Sichuan province, discovered in 1978, with an estimated 52 tonnes of gold reserves.[49] However, at present it remains unclear to what extent these reserves will prove economically feasible to exploit.

Future plans also include exploitation of bedrock gold resources in Chamdo and Shigatse prefectures, TAR, potentially more profitable than the region's alluvial mines, but requiring larger-scale operations. At the time that the Specialist Plan was written (mid-1990s), work on these resources was in the early exploration stages. The deposits that had been mapped in Chamdo appear to be of very low grade, and as such will probably only be exploited as a by-product of the Yulong copper mine (see page 31).

Mine worker (identity obscured), TAR, mid-1990s **© TIN**

47 They are located in the following counties: Thewo (Ch: Diebu), Luchu (Ch: Luqu) and Machu (Ch: Maqu) counties in Kanlho (Ch: Gannan) Tibetan Autonomous Prefecture (TAP); Zungchu (Ch: Songpan), Namphing (Ch: Nanping/Jiuzhaigou), Dzoige (Ch: Ruo'ergai) and Dzamthang (Ch: Rangtang) counties in Ngaba (Ch: Aba) Qiang and Tibetan Autonomous Prefecture; and Lithang (Ch: Litang) county in Kardze TAP

48 Taihe Zhou, Richard J. Goldfarb and G. Neil Phillips, 'Tectonics and distribution of gold deposits in China – an overview', Mineralium Deposita, 17 January 2002; Jingwen Mao, Yumin Qiu, Richard J. Goldfarb, Zhaochong Zhang, Steve Garwin, and Ren Fengshou, 'Geology, distribution, and classification of gold deposits in the western Qinling belt, central China', Mineralium Deposita, 29 January 2002.

49 Jingwen Mao et al., ibid

Copper mining – opening new opportunities for exploitation

> *"Copper is a greatly superior mineral in Tibet. With the development of modern industry, our country [China]'s shortage of copper resources has been increasing by the day. Therefore developing Tibet's copper deposits has real significance for reducing wastage of our country's capital reserves, and increasing income for the Tibetan economy".*
> TAR Specialist Plan[50]

Copper is vital to China's development and industrialisation, but as a raw material copper is in very short supply. The metal's main commercial value is based on its electrical conductivity and a major proportion of the world's commercially produced copper is used by the electrical industries in building construction, telecommunications, electronics and electronic products.[51] With China's current industrial development, urbanisation and expansion of telecommunications and the authorities' hopes to establish high-tech industry bases, copper is a valuable commodity. However, copper ore is one of the metal minerals in seriously short supply in China, along with other metals such as high-grade iron ore, chromite, cobalt, platinum and manganese. China only has a 40 per cent degree of self-sufficiency in production from its copper mines and the state has had to rely on increasing imports as demand increases.

Although China's copper mine output has risen over the past decade, consumption has grown at an even faster rate. China's copper mine output increased from an estimated 285,000 tonnes in 1990[52] to approximately 620,000 tonnes in 2001.[53] However China's domestic smelters remain heavily reliant on imports of non-refined copper (copper concentrate and copper scrap) in order to meet production targets for refined copper. The shortage is at the mine head; China does not have a problem meeting demands for refined copper. In fact in recent years, following the construction of new smelters and a dramatic boost in refined copper production, the price of copper in China has actually been pushed down by a surplus of refined copper on the domestic market. This boost in production has further increased the demand for copper ore and concentrates and it has only been possible to meet this demand

50 Op cit, p.502

51 According to the United States Geological Survey (USGS), the electrical uses of copper, including power transmission and generation, building wiring, telecommunications, and electrical and electronic products, account for about three quarters of total copper use. Building construction is the single largest market, followed by electronics and electronic products, transportation, industrial machinery, and consumer and general products. 'Copper: Statistics and Information', USGS website www.usgs.gov

52 Pui-Kwan Tse, 'The Mineral Industry of China, 1994', USGS, 1995. The figures here refer to tonnage of copper (Cu) content of ore.

53 'Copper', US Geological Survey, Mineral Commodity Summaries, January 2002 (www.usgs.gov). China's 2000 copper mine output was estimated to be 590,000 tonnes (Pui-Kwan Tse, 'The Mineral Industry of China, 2000', USGS, 2001). Chile is the world's largest copper producer, with mine output of 4,650,000 tonnes in 2001.

through using copper that is not of Chinese origin. The total value of copper imports in 2000 was US$5.3 billion,[54] as compared to only US$627 million worth of exports.[55] As a result of the imbalance between supply and demand and the cost of imports, China needs to increase domestic copper mine production. Mines that opened in the 1950s and 1960s are now nearing exhaustion. China is looking to its western regions to make up at least some of the shortfall by opening up known deposits that have not yet been developed and by pushing forward with exploration work to discover new reserves.

Copper processing

The processing and refining of copper involves several stages; it is at the mine head where the raw ore is extracted that China faces a fundamental shortage, meaning that in other stages of production imports are having to be relied upon. The copper content (Cu content) of the ore that is mined from most deposits is very low – often less than one per cent. The mined ore has to pass through several stages of processing and refining in order to produce pure copper (100 per cent Cu content).

Copper is usually found associated with sulphur. After the copper sulphide ore has been mined, it is crushed and ground, then milled into fine particles before being put through the flotation process, which involves the removal of sulphide particles through oxidation. The resulting solution (known as 'copper concentrate') is about 25-35 per cent copper content. The copper concentrate is then transported to a smelter where it is roasted and smelted with iron oxide to produce a molten sulphide called 'copper matte'. The copper matte is converted into 'blister copper' which is 98-99% pure through removal of the iron and further oxidation. The blister copper is put through a further process of fire refining and is then cast into thin sheets before being transported (as 'copper anode') to an electrolytic refinery, where remaining impurities are removed.

In addition to the smelting of sulphide associated copper, an increasing share of copper is now being produced from acid leaching of oxidized ores.[56] This process involves the passing of a solvent (sulphuric acid) through the copper-oxide ore to remove the copper. The resulting solution is taken to an extracting plant where the copper is extracted and further refined before being purified by an electrical process (electrowinning, known as the SX-EX process).

54 This includes ore, metal and alloys (unwrought), semi-manufactures, and scrap.
55 Figures from the General Administration of Customs of the PRC, 'China monthly imports and exports', no.12, 2000 cited in Pui-Kwan Tse 'The Mineral Industry of China, 2000', USGS, 2001, Table 4. According to Pui-Kwan Tse, most imported copper was from Australia, Chile and Mongolia.
56 'Copper: Statistics and Information', USGS, op cit.

Copper mining in Tibetan areas

After Jiangxi province on China's eastern seaboard, the TAR has the largest copper reserves in China concentrated in the east of the region in Chamdo prefecture. This zone contains the potentially 'world-class' Yulong deposit – the most significant of Tibet's verified mineral deposits outside the Tsaidam Basin.

Until now most of Tibet's copper, of great potential value to the state, has not been exploited. Unlike gold, copper is an expensive mineral to exploit and needs a medium to large-scale enterprise to carry out mining operations. Most of Tibet's copper, including the Yulong deposit, is of the porphyry type, meaning that the copper is scattered through the deposit resulting in low concentrations of copper in the rock.[57] For example, according to the TAR Specialist Plan, the average ore grade in the region is 0.35–0.9 per cent, common for porphyry type deposits. Only the large volume of ore makes it economic to mine such low grades and it is therefore necessary for these mines to be large-scale, extracting hundreds of tonnes of rocks per day in order to produce a profitable amount of processed or refined copper each year. A large-scale mine will process thousands of tonnes per day. Equipment and processing facilities also require huge investment and a solid infrastructure (transportation, communications, and power) is essential. Investment is substantial and generally for long-term rather than short-term turnover of profit and as a result there has not been the small-scale private mining of copper that occurs where gold is found. Due to a combination of the above factors, most copper deposits in Tibetan areas have not yet been developed.

However, the determination of the authorities to develop domestic copper sources and the large-scale investment in infrastructure construction in Tibetan areas are making these deposits increasingly accessible. Preparatory infrastructure construction (roads, housing, power) is reported to be underway for the development two of Tibet's copper deposits, including Yulong. The Western Mining Company, based in Qinghai province's capital Xining, has been awarded the contract to develop both mines.[58] The fact that a Xining-based company has been awarded the contract to develop Yulong rather than a TAR company could have an impact on the profitability of Yulong to the TAR's economy.

57 Porphyry: igneous rock (ie rock produced by volcanic or magmatic action) with large crystals scattered in a matrix of much smaller crystals (Oxford English Dictionary).

58 Pui-Kwan Tse, 'The Mineral Industry of China, 2000', USGS, 2001. According to this report, the Western Mining Company was formerly called the Xitieshan Mining Bureau and was located in Xitieshan, Tsonub M&TAP. It changed its name in 2001 and the authorities approved its re-location to Xining.

Yulong

The eastern area of traditional Tibet, known as Kham,[59] holds the largest known Tibetan copper reserves, with a rich porphyry copper belt running south through Chamdo prefecture (TAR) into Dechen TAP (Yunnan) and east into the edge of Kardze TAP (Sichuan). The largest known deposit in this copper belt is Yulong, located in Chunyido (Ch: Qingnitong) village, Jomda (Ch: Jiangda) county, Chamdo, in the east of the TAR. Yulong is estimated to have 6.5 million tonnes of prospective reserves, representing over one tenth of China's estimated total copper resources (62.74 million tonnes[60]). The potential value of Yulong was underlined when it attracted the interest of foreign mining companies in the mid-1990s, notably the US-based Cyprus Amax, who considered it to be the only mineral deposit in Tibet large enough to be of potential interest.[61]

The exploitation of Yulong was delayed during the 1990s by its remoteness and weak supporting infrastructure,[62] but development of the mine now appears to be going ahead. Local infrastructure construction has been proceeding, including the development of roads and hydropower resources,[63] and construction of housing for mine workers was reported to have commenced in summer 2001.

The original aim for Yulong in its first phase was to be producing 20,000 tonnes per year of electrolytic copper by 2000, to be increased to 100,000 tonnes by 2010.[64] Although development is behind schedule planned production capacity remains the same. An additional 100,000 tonnes of copper per year from one mine will provide a considerable boost to China's domestic copper mine production figures, representing nearly one sixth of current output. The plans for Yulong to produce electrolytic (refined, almost pure) copper rather than copper concentrate significantly adds to the value of the mine. According to the TAR Specialist Plan, a portion of the electrolytic copper will also be directly processed into copper products, further boosting the mine's profitability within the TAR and provide development of associated local industry.

59 See note 32.
60 Statistics from China Mineral Resources Report 1996-1998 and China Geology and Mineral Resource Yearbook 1991-1997 cited in Pui Kwan Tse, 'The Mineral Industry of China, 1999', USGS, 2000. According to this report, China has 25.25 million tonnes of copper reserves (i.e. resources that are considered feasible to exploit). The US Geological Survey estimates that China has 18 million tonnes of copper reserves and a reserve base of 37 million tonnes; it gives no figure for total resources ('Copper', USGS, January 2002, op cit). One of the problems in assessing statistical information on resources is the unreliability of geological data and the different classifications used for reserves and resources. See chapter 2, pp. 46-49.
61 Cyprus Amax withdrew their interest in the project due to unwillingness on the part of the authorities to relinquish their own interests in the project. See chapter 2, p. 69
62 See chapter 2, pp. 41-42
63 See chapter 2, p. 43
64 TAR Specialist Plan, op cit, p 125-6.

Industrial centres and key copper deposits in Chamdo prefecture

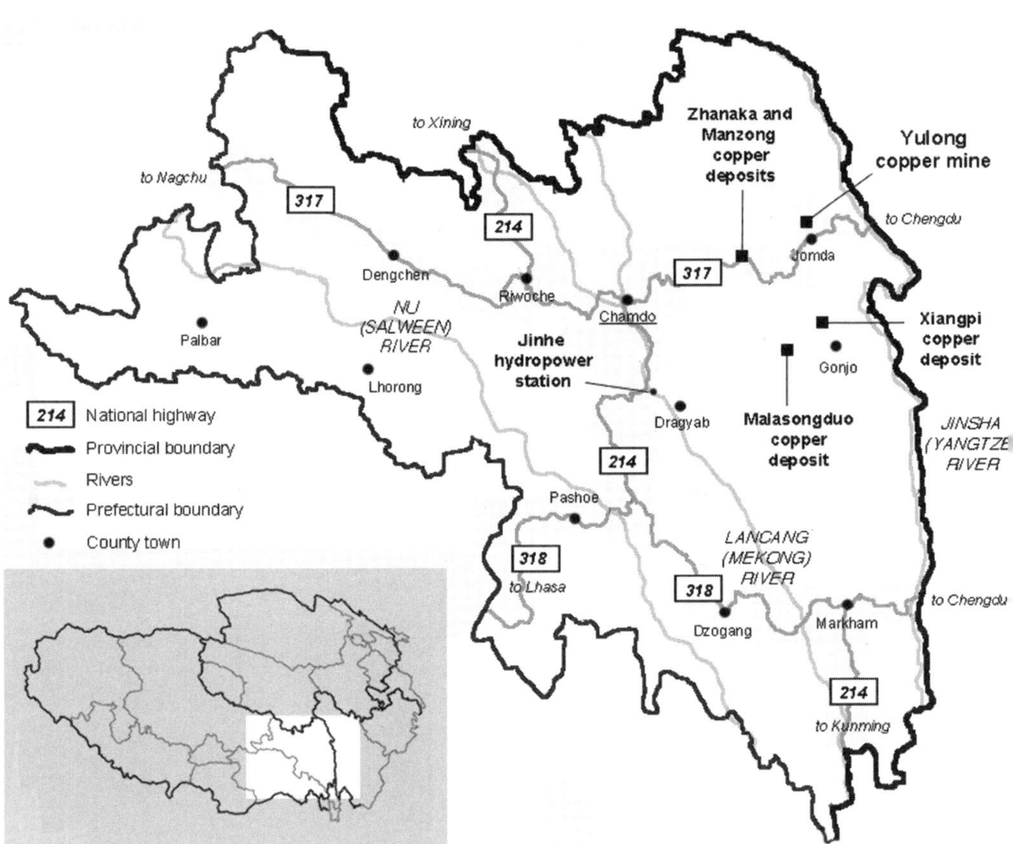

The Yulong copper mine is located in, what the TAR Specialist Plan refers to as, the 'Yulong mining zone', stretching for 400km through Chamdo prefecture, varying between 30 and 70km in width, with a total area of 1870 square kilometres.[65] The TAR Specialist Plan forecasts that the Yulong mine should have a life of 28 years, and mentions the need to develop other mining areas in the zone, identifying four other proven deposits to be prioritised for supplementary prospecting and development during the period 2000-2010.[66]

Porphyry copper deposits usually fall into two classes, one containing copper and gold and the other containing copper and molybdenum. The primary associated mineral of the deposits in the Yulong mining zone is molybdenum – making the deposits less profitable than if the primary associated mineral had been gold.

Saishitang

The other copper mine that looks set to become operational in the near future is the Saishitang copper mine in Tsigorthang (Ch: Xinghai) county Tsolho (Ch: Hainan) TAP, Qinghai province. Tsolho contains a reported 62 per cent of the Qinghai province's copper reserves,[67] and Saishitang appears to be its most potentially valuable mine and one of the three largest verified copper deposits in Qinghai. According to a Tibetan who used to work in a government administrative department in Qinghai, Chinese workers arrived at the Saishitang mine site in summer 2001 to prepare infrastructure (roads, power, housing) for the mine. The Qinghai Province Land and Resources Bureau have reported that a mining and dressing plant should be opened at Saishitang in 2003, designed to process 500,000 tonnes of rock per year and forecast to produce 6,580 tonnes (copper content) per year of copper concentrate.[68] Another of Qinghai's three main copper deposits, not yet being developed, is Tongyugou, located in the same copper belt (the Saishitang-Tongyugou-Suolagou copper belt). Both of these mines, discovered before 1986, are reported to have high copper concentrations (1.25% and 1.23% Cu content respectively)[69] and large reserves, although nothing like the scale of Yulong.[70]

65 Also referred to as the Yulong bornite zone. Bornite is a copper mineral (copper-iron sulphide).
66 Place names given in Chinese pinyin: Malasongduo, in Kuoda village, Chaya (Tib: Dragyab) county; Duoxiasongduo, in Seduo village, Xiangpi township, Gongjue (Tib: Gonjo) county; Zhanaka and Manzong mining areas in Tuoba township, Changdu (Tib: Chamdo) county. TAR Specialist Plan, op cit, p. 495-500.
67 Tsolho prefectural government web pages on Xining's official website www.xining.gov.cn
68 Just under 20 per cent of the forecast production of 39,200 tonnes per annum of ore with a concentrated copper content of 20 per cent. Where copper concentrate production tonnage is listed this refers to the amount of copper (the Cu content) of material produced, not the total tonnage of ore (which would be 70-80% non-copper).
69 'Introduction to E'la mountain region established gold, copper and other metal minerals belt', Qinghai Land and Resources Bureau website www.qhgtzy.gov.cn.
70 According to Qinghai province land and resources bureau Saishitang has 417,000 tonnes of verified [*tan ming*] B, C and D grade resources, while Tongyugou possesses guaranteed [*kong zhi*] D grade reserves of 500,000 tonnes (ibid).

Tsolho TAP has the advantages of relatively good transport and communications lines compared to the TAR and much of the rest of Qinghai. The prefecture shares a border with Xining municipality to the north, which is linked to the rest of China by rail.[71] Current infrastructure construction in the area includes a four-lane expressway and a railway line.[72] The Yellow River (Tib: Machu; Ch: Huanghe) runs through the prefecture, providing important power sources from the two large-scale dams at Longyangxia and Lijiaxia.

By comparison, the other large copper mine in Qinghai is De'erni in Dawu township, Machen (Ch: Maqin) county, Golog (Guoluo) TAP, is much more remote and with much weaker infrastructure support than Tsolho. As a result it is likely to take longer for exploitation to become feasible. De'erni is reported to have high grade copper reserves and also contains large reserves of cobalt (a mineral in short supply in China), gold, sulphur and selenium and medium sized zinc and silver deposits.[73] The mine is not yet operational but has set the same target as the Saishitang mine for its first phase of development.

Working mines and other deposits

Copper is mined at the multi-metals mine in Gyama Trikhang, Meldrogongkar (Ch: Mozhugongka) county, Lhasa muncipality. The mine, which was officially opened in 1993, was one of the TAR Specialist Plan's seven key 'priority construction projects' to be focused on before the year 2000. The aim was to produce 3,750 tonnes per year of copper concentrate by the year 2000, as well as 11,250 tonnes of lead concentrate and 13,750 tonnes of zinc concentrate. The Gyama mine remains a priority of the TAR authorities; it is listed as one of three 'major development projects', alongside Yulong copper mine and the Zhabuye salt lake in the TAR's Tenth Five-Year Plan (2001-2005).

71 Tsolho originally bordered Haidong prefecture, however, two of Haidong's counties were transferred to Xining in 2000. See 'County transfers facilitate industrial development in Qinghai', TIN News Update, 22 January 2002.

72 See 'New railway construction in Qinghai', TIN News Update, 15 January 2002.

73 Qinghai Land and Resources Bureau website qhgtzy.gov.cn. The De'erni copper deposit reportedly contains various mining sites with total verified [*tanming*] reserves of 556,000 tonnes of copper, of which 338,000 are of A, B and C grade, with an average Cu content of 1.27 per cent. 'Introduction to De'erni mining area (*De'er ni kuang qu jian yi*)', Qinghai Province Land and Resources Bureau website www.qhgtzy.gov.cn

74 (see next page) Rebgong (Ch: Tongren) county gold factory in Malho (Ch: Huangnan) TAP in Qinghai used to process copper, but this operation was reportedly closed down during the late 1990s.

75 (see next page) Pui-Kwan Tse, 'The Mineral Industry of China, 1997', USGS, 1998. According to this report there were plans to extend the handling capacity to 800 tonnes of ore per day. According to the official Kardze Annals (1997), the mine first opened in 1985 as a county-level enterprise, but by 1990 had received state, provincial and prefectural investment to increase production. In 1990 Liwu reportedly made a profit of 4.78 million yuan, becoming 'one of the most important pillars in the county's income' (Kardze Annals, p.122-3). According to the Sichuan provincial government, Liwu possesses 260,700 tonnes of copper reserves at a high average ore grade of 2.5 per cent copper content (The Sichuan government's official website page on Jiulong county, www.sc.gov.cn).

76 Reported to have reserves of 20,200 tonnes at an average ore grade of 1.65 per cent copper content.

77 'Kardze Annals', 1997, op cit.

According to TIN research, Kardze TAP in western Sichuan is the only other Tibetan area with known working copper mines.[74] The largest is the Liwu copper mine and processing plant in Gyezur (Ch: Jiulong) county, that opened in 1985. By the end of the 1990s it was designed to handle 500 tonnes of ore a day of copper concentrate.and process 4,000 tonnes per year [75] Located south of Kardze prefecture's capital Dartsedo, Gyezur county town is roughly 200km by road from the Chengdu-Kunming railway. Most of Liwu's copper concentrates are reportedly being sold to the Yunnan smelter in Kunming, and will therefore be counted in Yunnan's refined copper production figures. Another operational copper mine in Gyezur is the much smaller Wajingou copper mine.[76] Payul and Dege counties in Kardze, bordering Jomda county in Chamdo, are also reported to have operational copper mines,[77] probably located in the same copper belt as Yulong. Copper is also mined in Dartsedo and Dabpa counties.[78]

Yunnan province is currently the second largest copper producer after Jiangxi, but the copper reserves in Dechen (Ch: Diqing) TAP, Yunnan, belonging to the same belt as Yulong, are not yet an important contributor to provincial copper output. However, as infrastructure and investment conditions improve, these reserves are likely to be exploited. The Yunnan provincial mining bureau is currently seeking domestic and foreign investment for further exploration and development of the Yangla copper mining zone in Dechen (Ch: Deqin) county, in the north-west of Yunnan bordering the TAR,[79] and Dechen county geology and minerals department is advertising for investment in a project to build a copper processing plant for Yangla with a daily capacity of 500 tonnes.

The opening of the Golmud-Lhasa railway will significantly improve the feasibility of exploiting copper and other minerals in Tibetan areas, particularly those near to the route the railway follows south through Yushu TAP in Qinghai into Nagchu prefecture in the TAR and on to Lhasa. There have been various recent reports of discoveries of copper deposits in the Lhasa valley, an area with the benefits of relatively developed transportation and power networks. There have also been several reports of new copper deposits lying alongside the route of the Golmud-Lhasa railway since its construction started in June 2001. For example, in Jaunary 2002 Xinhua reported that a 'large copper mine' had been found in Nyemo (Ch: Nimu) county, Lhasa municipality close to the railway. On the same day Xinhua also reported that 13 high grade copper deposits had been discovered around the Fenghuoshan Basin in Yushu, also on the railway route.[80]

74, 75, 76 & 77 See footnotes on previous page
78 Ibid.
79 'Exploration and development of Yangla copper ores in Deqin, Diqing Prefecture', project summary on the Yunnan Provincial Government Foreign Affairs Office's 'Investment in Yunnan' website (in English and Chinese) www.yunnantea.net/invest/index.shtml
80 Qinghai Daily, 7 January 2002

Proposed route of Golmud–Lhasa railway (under construction)

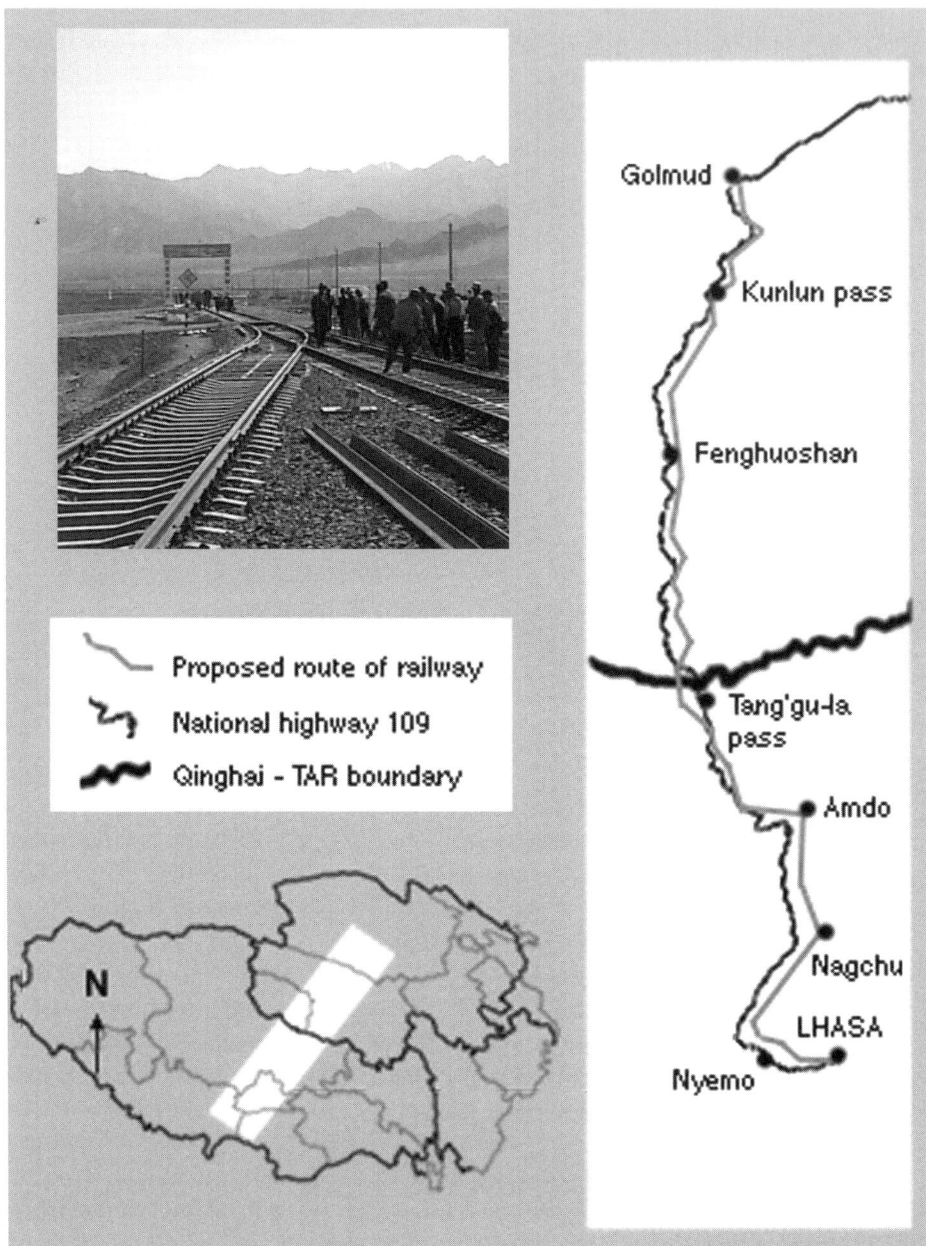

Insert: Railway workers at Nanshankou, south of Golmud, 2001

2. Obstacles to development

China's ambitious plans for developing Tibet's resources are based on the widely held perception of Tibet as a rich treasure house of minerals ripe for exploitation. However, there are many obstacles that the authorities need to overcome if they are going to succeed in meeting domestic demand for raw materials by exploiting Tibet's resources. Tibet is well known for its remoteness, high altitude and harsh climate. It is largely these conditions that have so far limited mineral exploitation. Other obstacles facing the authorities in the development of the mining industry in Tibetan regions, for example poor management and control of resources, inefficiency and corruption, are similar to problems facing the development and reform of the mining industry throughout China. However, these difficulties are exacerbated by Tibet's distance from markets and relatively undeveloped infrastructure. Despite the reform of the legislative system governing mining and the introduction of preferential policies to draw financial and technological investment to western China, attracting investment into the minerals industry in Tibet, particularly from larger foreign mining companies, remains a real challenge for the authorities. In addition, as China enters the global economy, it faces increasing competition in the minerals industry, which is marked at the moment by growing competitiveness and falling prices. By examining these problems, this chapter seeks to cut through the exaggerated reports of the economic potential for mining to provide a more realistic perspective on the past and future development of Tibet's minerals industry. It also looks at some of the steps that the Chinese authorities are taking to overcome these obstacles to development as a result of their unwavering determination to make the large-scale exploitation of Tibet's mineral resources feasible.

The first section looks at the technical issues of poor infrastructure, the lack of reliable geological data and the shortage of technology and skilled personnel. It also outlines current development policy and plans aimed at addressing these problems – for example, construction of the Golmud-Lhasa railway and the relaxation of restrictions on population movement and residency. The second section focuses on the poor management and regulation of resources, which when combined with the push for fast-track development has led to irrational resource exploitation. It considers the negative outcomes of this from the perspective of the central authorities, who have lost potential value of State resources through extensive rather than intensive mining practices and through illegal mining and trading. A case study on the development of the gold industry in China is provided as an example of the macro measures that the central authorities are taking to strengthen State control and to reform and restructure the mining industry in attempt to move from small-scale to large-scale exploitation.

The third section deals with the need to attract investment into the mining industry in Tibetan areas, focusing in particular on foreign investment. Providing an overview of the reform of the legislative and regulatory environment for investment, it also outlines some of the preferential policies developed to attract investment into Tibetan and other western areas of China. It looks at the limited extent to which foreign companies, increasingly interested in China's minerals industry, have already become involved in Tibetan areas. However, it is considered that, in the near future at least, investment in the exploitation of non-oil and gas minerals in Tibetan areas is likely to remain largely in the hands of the State and domestic private enterprises, with some investment from junior foreign mining companies.

NOTE: This chapter seeks to outline the obstacles to development and measures being taken to overcome these obstacles from the perspective of the Chinese state. The state has claimed ownership of all mineral resources within its territory, and is committed to exploiting Tibet's resources. For further discussion of the impact of these problems and policies on Tibetans – for example local protectionism or increased immigration of Chinese into Tibetan areas see Chapters 3 and 4, covering land and labour issues. See Chapter 6 for a discussion on the model of development being applied to Tibetan areas and the distribution of benefit from mining. It should also be noted here that one of the greatest challenges facing China is ensuring adequate protection of the fragile Tibetan environment. This issue is dealt with in Chapter 5.

Gold-mining machine frozen in riverbed in mid-winter, Lithang (Ch: Litang) county, Kardze (Ch: Ganzi) TAP, Sichuan

Technical problems

Infrastructure

> *"Tibet has a wealth of natural resources [...] all of which are ripe for exploitation, but, constricted by the condition of basic communications infrastructure, the level of exploitation and utilisation of these natural resources is extremely low".* Tibet Autonomous Region National Land Specialist Plan (1996-2020).[1]

The remoteness of Tibetan areas remains one of the main obstacles to the development of the mining industry and to attracting foreign capital and technological investment. The inhospitable climate presents many difficulties, not least the freezing temperatures in winter that make mining impossible for several months each year. In order to make a mining enterprise economically viable two main conditions need to be met: an adequate power supply and the ability to transport raw materials to processing centres and markets. If the Chinese authorities are to succeed in their large-scale exploitation plans development of a basic infrastructure is absolutely essential. The current drive to develop the western regions recognises this need for improved infrastructure for development, including mining, and is based on large-scale infrastructure construction, including railways, roads and power.

It is not a coincidence that the extractive industries of Tibetan areas lying along the Lanzhou-Xining-Golmud railway are more developed than in other Tibetan areas. The Xining-Golmud section of the railway was opened in 1984. Heading west from Xining, the railway skirts the northern shores of Qinghai Lake (Mong: Kokonor; Ch: Qinghai Hu), where a branch line heads north from Ha'ergai to the Reshui Coal Mine in Tsojang (Ch: Haibei) Tibetan Autonomous Prefecture (TAP). Further west another branch line heads south from Chahanruo to Tsakha (Ch: Chaka) salt lake in Wulan county, Tsonub (Ch: Haixi) Mongolian and Tibetan Autonomous Prefecture (M&TAP).[2] The railway continues west through the prefectural capital Terlenkha (Ch: Delingha), before turning south-west just south of the Dachaidan mining area,[3] and passing through

1 '*Xizang Zizhiqu Guotu Zhuanti Guihua* (1996-2020)', a two volume internal (*neibu*) document formulated by the TAR National Land Plan Formulation Committee (*guotu guihua bianzhi weiyuanhui*) and the TAR Planned Economy Committee (*jihua jingji weiyuan hui*). Although undated, the document was written after the Third Tibet Work Forum (1994) and is believed to have been issued 1994-1995. Referred to below as the 'TAR Specialist Plan'.

2 There is a salt mine and factory at Tsakha. See Chapter 3, p.85. Forced immigrant settlements and agricultural laogai were established at Wulan's county seat and in the Chaka district during the 1950s. Marshall and Cooke, 1997, op cit.

3 Dachaidan is one of three specially designated county-level administrative committees (*xingzheng weiyuanhui*) in Tsonub TAP (Qinghai Statistical Yearbook, 2001). Construction of a settlement at Dachaidan began in 1956. Dachaidan has jurisdiction over the townships of Dachaidan, Xitieshan and Yuka (Tsonub/Haixi prefecture website www.haixidxj.com.cn). Mineral resources in the area include salt, oil and gas and gold. Lead and zinc has been mined at Xitieshan since 1957 (Marhsall and Cooke, 1997, p.1862). The other two county-level administrative committees in Tsonub are Lenghu (see note 9) and Mangya (see note 8).

the Cha'erhan mining area[4] into Golmud.[5] National Highway 315 follows the railway route, while National Highway 109 passes south of Qinghai Lake and through Dulan county.[6] National Highways 215 (to north-western Gansu) and 315 (to Xinjiang), and a road linking 215 and 315 provide a link to the railway from the other major mining areas of Tsonub: Dachaidan, Mangya[7] and Lenghu[8]. Further expansion of this transportation network was reported in January 2002, when TIN received an eyewitness report of the construction of a single-track railway, heading south from Tongkor (Ch: Huangyuan) county town (on the main rail line) towards Tsolho (Ch: Hainan) TAP. Although the final destination of the railway is unknown, any rail connection with Tsolho will facilitate the exploitation of the prefecture's copper resources, amongst other development opportunities.[9]

The Tibet Autonomous Region (TAR), in contrast, is the only region of China not yet linked to the national railway network and remains relatively undeveloped and remote from the rest of the country. The TAR National Land Specialist Plan, an internal (neibu) document detailing the region's plans for the development of resources over the period 1996-2020, refers to the geographical situation of the TAR and its lack of adequate infrastructure for economical mineral exploitation:

4 Cha'erhan's industry is based on its salt lake resources, including potassium, an important raw material for potash fertiliser. The Qinghai Salt Industry Corporation claims that the Cha'erhan salt lake has the largest soluble potassium-magnesium deposit in China and one of the largest in the world. See An Pingshui (Qinghai Salt Industry Corporation), 'Introduction to Qinghai Potash Fertiliser Project', International Fertiliser Industry Association (IFA), 2000, full text published on IFA website, www.fertilizer.org . According to this report, Cha'erhan's resources have been exploited since 1958, but not on a large-scale until 1982 when the Qinghai Potash Fertiliser Plant was established. In 1997, the Qinghai Salt Lake Potash Fertilizer Co (an amalgamation of Qinghai Salt Industry Corporation and six other enterprises, including Beijing Power Industrial Company) issued stocks on the Shenzhen stock exchange. Cha'erhan was previously a county-level administrative committee, but is not listed as such in the 2001 Qinghai Statistical Yearbook and has probably been incorporated into Golmud municipality or Dachaidan. Two other areas, Kunlun and Hanhai, also appear to have lost their county-level status and have probably been incorporated into Golmud.

5 Golmud is the main urban and industrial centre serving Cha'erhan, Dachaidan, Lenghu and Mangya. Golmud municipality (Ch: *shi*; county level) is also responsible for administering a large zone in the south-west corner of Yushu prefecture.

6 Agricultural laogai were established in Dulan county during the 1950s to support the development of the mining industry to the north and west. The Xiangride prison farm, formally opened in 1956, still operates as a labour camp and is very important to Tsonub prefecture's grain production (James D. Seymour and Richard Anderson, 'New Ghosts, Old Ghosts' , New York and London: M.E.Sharpe, 1998, p.139). Dulan itself has rich mineral reserves, particularly iron, although they do not appear to have been developed much yet.

7 Prisoners and forced immigrants began construction on a settlement at Mangya in 1956 (Marhsall and Cooke, 1997, op cit). A temporary work committee was established in Mangya in 1958 to exploit the asbestos and oil resources in the area. It was returned to the jurisdiction of Tsonub M&TAP as a township from the early 1960s until the establishment of the county-level Mangya Administrative Committee in June 1984. (www.haixidxj.com.cn). According to the China National Non-Metallic Minerals Industry Corporation (CNMC), the Mangya Asbestos Mine, founded in 1958, has the largest proved reserves in the country and is at present able to produce 80,000 tonnes of asbestos per year (www.cnmc.com).

8 Lenghu Administrative Committee (county-level). Extensive oil reserves were discovered at Lenghu in 1959 and according to Marshall and Cooke these became one of the principal petroleum resources within Chinese-ruled territories, and of great importance in supplying the development needs of the new bases for settlement and exploitation in Amdo and Central Tibetan regions (Marhsall and Cooke, 1997, op cit), p.1862

9 See Chapter 1, p.33 and 'New railway construction in Qinghai', TIN News Update, 15 January 2002.

Infrastructure and centres of mining activity in North Qinghai

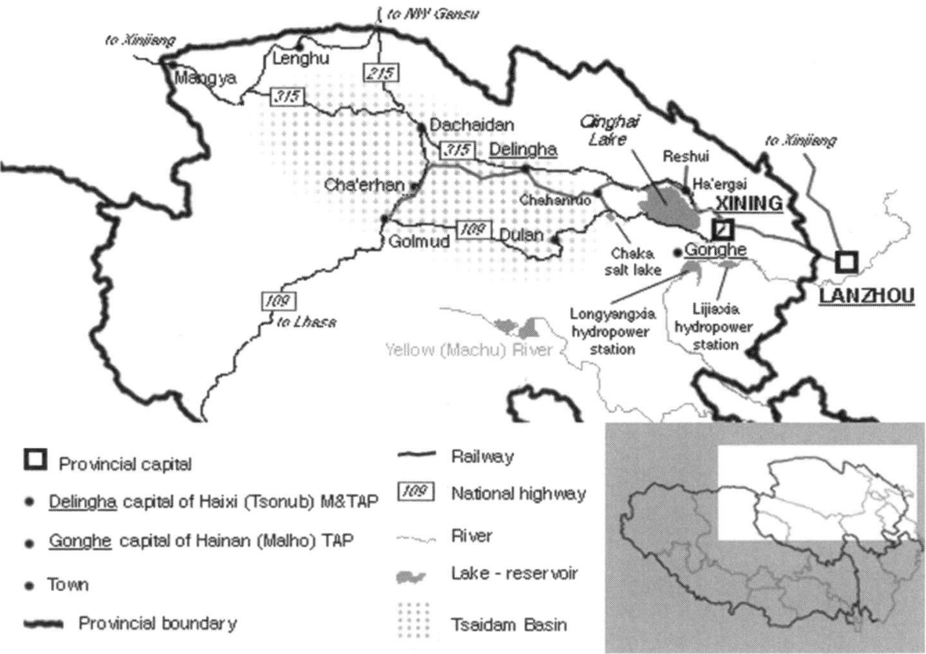

Place names given in Chinese pinyin

> *"Tibet is on the south-western frontier of the motherland. Its greatest characteristics are that the terrain is high and the climate is cold, the air is rarefied, and communications are inconvenient. Many mines are far from trunk highways and so there are mines for which it is hard to transport the material out. Even if it is being transported to the interior [neidi], transport costs are too high and so the mining products lack competitiveness".*[10]

The remoteness of the region has meant that the authorities have had problems developing even the most promising of the TAR's reserves – the Yulong copper mine in Chamdo (Ch: Changdu) prefecture, which is reported to be a 'world-class' deposit.[11]

10 TAR Specialist Plan, see note 1 for details.

11 The Yulong copper mine, located in Chunyido (Ch: Qingnitong) village, Jomda (Ch: Jiangda) county, Chamdo (Ch: Changdu) prefecture, is the largest known deposit in what the Specialist Plan refers to as the 'Yulong copper mining zone'. The zone has a total area of approximately 1870 square km (400km long and varying between 30 and 70km wide) within Chamdo prefecture, and is part of a copper belt that stretches south into Yunnan province (see map p.32 and also main map, inside back cover).

The two main obstacles to exploitation are transportation and energy supply. The Specialist Plan had set a target for the Yulong copper mine to be in operation by the year 2000 with an annual production capacity of 20,000 tonnes of electrolytic copper. However, development of the deposit was delayed. In 2002, TibetGuide.com, a website that lists investment opportunities in TAR projects, was still listing Yulong as requiring investment of US$144.58 million for the first phase of exploitation (annual production capacity 20,000 tonnes). The project information, which appears to be the same as that used at a TAR investment and trade seminar in October 2000, states that: "*There have already been plans as to how to resolve the energy resource and transportation problems*". These include renovation of the northern Sichuan-Tibet highway (National Highway 317) that passes through Chamdo, and development of the Jinhe hydropower station.[12] By summer 2001, preparatory work (housing and road construction) was proceeding at the mine site and the authorities had reportedly marked off the mining area. Work will have stopped during the winter months, and it is not yet known if the mine has become operational in summer 2002.

The construction of the Golmud-Lhasa railway, one of China's Tenth Five-Year Plan's 'priority projects', will have a major impact, if completed successfully, on the viability of exploiting the mineral resources of the remoter areas of Qinghai along its route, and those of the TAR. Since construction started, the Chinese press has announced the discovery of a large copper belt at Fenghuoshan in Yushu TAP, which will be one of the stops on the new railway line (see map p.36). Thirty mining projects are also reportedly being undertaken along the route of the railway during the plan period (2001-2005).[13] The railway, due to be finished by the end of the plan period, will also increase the feasibility of exploiting known resources in other parts of the TAR.

Considerable attention is also being paid to road building and renovation of existing state highways. For example, during the 10th Five-Year Plan period (2001-2005), the TAR will spend 12 billion yuan (about US$1.45 billion) on 51 highway construction and technology projects.[14] A new four-lane expressway is under construction in Qinghai province that will ultimately link Lanzhou (capital of Gansu) to Lhasa via Xining and Golmud. It follows the route of the existing National Highway 109, but will be able to handle significantly higher traffic volumes at greater speed.[15] On the southern Sichuan-Tibet highway (318), a new tunnel was opened in autumn 2000 cutting through Erlangshan (Erlang Mountain), significantly improving links from Sichuan's capital Chengdu to the Tibetan areas lying to the west (Kardze/Ganzi TAP and beyond that the TAR). Previously the road took a much longer and more arduous route over the Erlang pass.

12 See below.
13 China's Tibet Information Centre, www.tibetinfor.com
14 Xinhua, 9 April 2002
15 Some of the construction is new, while other sections of the existing highway are being widened and paved. See 'New railway construction in Qinghai', TIN News Update, 15 January 2002

China is also seeking to develop the hydropower resources of the Tibetan plateau. This will support industry in the east as well as the development of the western regions.[16] The mining industry is a huge energy consumer, involving as it does the movement and processing of large volumes of material.[17] Traditionally coal has been the principal fuel used for power generation in the mining industry. However, Tibet is reported to have fairly low grade coal reserves and given increasing global concerns over CO^2 emissions and global warming alternative power sources are being sought. The Tibetan plateau has abundant water resources, including the headwaters of many of Asia's major rivers. The quantity of water flowing from the plateau and the steep gradients mean that the hydropower potential of Tibet's rivers is among the highest in the world and constitutes a very high proportion of China's overall hydropower potential.[18] Two of China's largest hydropower plants are located on the upper reaches of the Yellow River in Tsolho TAP, Qinghai. Lijiaxia, in the east of Tsolho bordering Malho (Ch: Huangnan) TAP and Tsoshar (Ch: Haidong) prefecture, comes under the direct jurisdiction of the province, while Longyangxia, near Chabcha (Ch: Gonghe), is a county-level unit.[19] The Saishitang copper mine, currently being developed in Tsolho, is expected to be powered by the dams on this stretch of the Yellow River.

Plans to construct new stations and expand existing plants include work on the Jinhe power station on the Mekong river to power development of the Yulong copper deposit in Chamdo. The power station is located in Dragyab (Ch: Chaya) county, 53km to the south east of Chamdo prefecture's capital (Chamdo county town). Chamdo's power grid was renovated in 1994-1995 as one of the Third Tibet Work Forum's 62 Projects[20] and construction is currently underway to build a 60,000 kW capacity generating set for the mine in order to "*alleviate the severe electricity shortage in this prefecture, and create good conditions for the development of the Yulong Copper Mine*".[21] The project, supervised at regional level, was being offered as a joint venture with the TAR Power Company; investment of US$53.7 million was being sought. However, construction had started on the power station by the end of 2001, and it has been referred to as a "*major investment project during Tibet's 10th Five Year Plan*".[22] This suggests that the project has gone ahead with state funding.

16 For example, one of the priority projects of the Tenth Five-Year Plan is to transfer electricity (generated by hydropower) from west to east China.

17 Even though the minerals sector is relatively small in the global economy, it is sometimes said that 4-7% of global energy demand is used in 'mining'. However, it is very hard to determine an accurate figure. ('Mining, Minerals and Sustainable Development Project Draft Report for Comment: Executive Summary', IIED, 4 March 2002)

18 'Tibet 2000: Environment and Development Issues', Environment and Development Desk, DIIR, 2000

19 Xue Zheng (ed), Qinghai Province Bureau of Statistics, 'Qinghai Statistical Yearbook, 2001', China Statistics Press, Beijing 2002.

20 See Chapter 6, p.187

21 Jinhe power station project summary on Tibet Guide www.tibetguide.com.cn.

22 'Tibet Starts Building Jinhe Power Station', Xinhua, 10 December 2001. An official opening ceremony was held on 6 December 2001

Dam construction in Lhoka (Ch: Shannan) prefecture, TAR, 1998 © TIN

The new facility is expected to go into operation in 2004. Although Jinhe's capacity will be sufficient for mining, if refining is to be done close by, as suggested by the TAR Specialist Plan, greater power capacity will probably be needed.

Much of the financing for new infrastructure projects (transportation, power, water and telecommunications) to be carried out as part of the Western Development Campaign comes from the central budget or from bank loans. China also hopes to attract private investment, including foreign investment, into infrastructure construction by offering preferential policies such as corporate tax cuts or exemptions[23] and relaxation of restrictions governing the length of allowed proportions of foreign equity and concessional foreign loans in total project investment.[24] Poverty alleviation funds will also be used in infrastructure construction.[25]

23 For example, domestic enterprises will be exempt from tax for the first two years of production, and will only pay 50 per cent of tax for years three to five. Foreign investors are offered slightly less favourable terms: those who will stay in business for more than ten years will be exempt for two years following the year that they first make a profit and will only pay 50 per cent for years three to five. 'Implementation Opinions Concerning Policies and Measure Pertaining to the Development of the Western Region', dated 28 August 2001, published by Xinhua in Chinese, 20 December 2001. Translated and published by FBIS in English 15 January 2002. Referred to below as the 'Implementation Opinions'.

24 See for example Articles 42-44 of the 'Implementation Opinions', op cit.

25 Article 8, 'Implementation Opinions', op cit.

However much is invested in improving infrastructure, the fact remains that Tibetan areas are a long way from the eastern seaboard and the cost of transportation to the market place will remain relatively high. As China meets the agreements set with its accession to the WTO, mining enterprises will also have to compete with cheap imports of raw materials. Smaller-scale local government and private enterprises at least will find it hard to compete.

One way in which the infrastructure problems can be eased somewhat is to develop the refining and processing industries in Tibetan areas and to cut down on transportation needs by creating an industry based on value-added semi-finished and finished products. For example, the TAR Specialist Plan's targets for developing the TAR's chromite industry are based around production at the Norbusa Chromite mine in Chusum (Ch: Qusorg) county, Lhoka (Ch: Shannan) prefecture – China's largest chromite mine.[26] Chromite is used to harden iron, steel (for example in the production of stainless steel), and nonferrous alloys and also increases their resistance to corrosion and oxidation. The Plan states that: 'at present, Tibet's chromite is sold in the form of raw ore, and lengthy transportation distances have had a direct impact on economic returns from the industry. However, the Plan states that the iron alloy factory in Lhoka (near Norbusa) and the machinery repairs factory in Lhasa have both 'basically grasped' the technology to produce ferrochrome alloys.[27] The Plan concludes:

> *"Consequently, at the same time as continuing to expand the scale of chromite production, [we will] accelerate the pace of production construction [shengchan jianshe] of ferrochrome alloys and gradually change the focus from the sale of ore to the sale of ferrochrome alloys."*[28]

However, the development of processing industries in Tibetan areas could be limited depending on the demand for raw materials from established industries in the central and eastern regions of China. The Tenth Five-Year Plan states that the processing of Tibet's mineral resources in the interior should be encouraged, indicating that the majority of mineral resources exploited in the TAR at least will continue to leave Tibet in the form of raw materials. The central authorities are also encouraging enterprises in the western regions to invest and build factories in neighbouring countries to process products and add value to raw materials.[29]

26 For more details on the Norbusa Chromite Mine see p.117

27 An alloy of iron and chromite with up to 72 per cent chromite. Ferrochrome is commonly used as a raw material in the making of stainless steel (The Steel Glossary, American Iron and Steel Institute, 2000, www.steel.org).

28 TAR Specialist Plan, op cit.

29 TAR Tenth Five-Year Plan; Implementation Opinions, op cit. The distribution of benefits from mining, including the removal of raw materials from Tibetan areas, will be discussed in Chapter 6.

Geological data and feasibility

> *"The level of geological work in Tibet has been low, and the geological results presented are lacking in reliability". TAR Specialist Plan.*[30]

In order to exploit mineral resources in an organised, efficient and profitable way it is essential to have reliable geological data and feasibility assessments. The mining industry in China, as in many developing countries, suffers from limited and unreliable geological data. Large areas in the remoter western regions in particular remain unmapped and unexplored. Most of the data that do exist are considered unreliable. At an international conference in October 2000, vice-minister for Land and Resources Shou Jiahua referred generally to the low level of geological exploration work in western China.[31]

In recent years China has pushed forward with geological exploration while improved infrastructure and access to better vehicles have made Tibetan areas increasingly accessible to surveying teams. The State Council has stipulated that exploration in China's western regions is to be given priority and will be favoured in the distribution of funding for geological surveying.[32] In 1999, the Ministry of Land and Resources invested 12bn yuan (US$1.4bn) in a 12-year mineral exploration programme, dubbed the 'Great Exploration of Land and Resources' by the Chinese press, which appears to be focused on mapping the natural resources of the western regions of China. In particular it will involve the surveying of 1.1m square km of land that hasn't been surveyed, most of which is on the Tibetan plateau. Discoveries so far reportedly include the above-mentioned new deposits along the route of the Golmud-Lhasa railway currently under construction.

There has also been some foreign involvement in recent attempts to develop mineral surveying. For example a recent collaborative project to comprehensively map China's gold resources, based on existing data and new exploration, has recently produced two detailed papers published in the academic journal Mineralium Deposita.[33] The authors include employees of the US-based Newmont Mining Corporation (one of the world's largest gold mining corporations) and several Australian juniors[34]

30 Op cit.

31 Shou Jiahua, 'Mining Development in Western China', paper given at the International Conference on Western Development and Mining Investment in China, Xinjiang Uighur Autonomous Region (XUAR) October 2000; papers from the conference can be accessed on the Ministry of Land and Resources website www.mlr.gov.cn

32 'Implementation Opinions', 2001, op cit.

33 Taihe Zhou, Richard J. Goldfarb and G. Neil Phillips, 'Tectonics and distribution of gold deposits in China – an overview', Mineralium Deposita, 17 January 2002; Jingwen Mao, Yumin Qiu, Richard J. Goldfarb, Zhaochong Zhang, Steve Garwin, and Ren Fengshou, 'Geology, distribution, and classification of gold deposits in the western Qinling belt, central China', Mineralium Deposita, 29 January 2002. See also Chapter 1, p. 27

34 A junior mining company is a small company that raises funds to pay for mineral exploration by selling shares on a stock exchange.

including Sino Mining Ltd., as well as the US Geological Survey and the University of Western Australia. The Chinese contributors are affiliated to Chinese academic institutions and the Gansu Bureau of Geology. One of the papers focuses on gold deposits in the western Qinling fold belt, that include 16 deposits in Tibetan areas of Gansu and Sichuan. There is also a long-term co-operative research project at the Geology Department of the University of Tasmania, involving a number of mining companies, including Sino Mining, who are investigating the mineral potential of East Asia, including China. The involvement of foreign mining companies, academics and geological surveys, could lead to improvements both in the quality of China's geological work and also in its credibility with mining companies.

However, up to the present, there is still a shortage of reliable data regarding most of the mineral deposits in Tibetan areas of the PRC. The TAR Specialist Plan stated that the geological results produced for the TAR by the mid-1990s were unreliable, partly because some of the data included in the statistics had not gone through examination and approval by the relevant competent bodies and concluded that:

> *"[Geological] work levels are low and the reliability of reserves is lacking, often causing government and investors to have no way of making decisions on whether or not to develop or to construct, or to make the wrong policy decisions."*[35]

This means that many mining enterprises in Tibet are operating without adequate data on the geology of a given area. In 1996, the general manager of the Nagchu (Ch: Naqu) Prefecture Mineral Development company published an article in the Tibet Daily outlining the problems of developing the mineral industry in northern TAR ('*Zangbei*'). This is one of the areas marked for exploitation of alluvial gold and antimony resources, and according to this article, chromite, copper, lead and zinc and coal are also being extracted. He pointed out that resources were not being exploited on a sound scientific basis:

> *"There is no prospecting data for a certain proportion of minerals being mined and production is organised only in accordance with experience and what can be seen on the surface of the land. Development prospects are not clear, and the development of enterprises will be adversely affected by insufficient reserves of natural resources. In addition, the area is vast, transportation routes are long and the period when production can be carried out is too short; the excessive cost of electricity means that the price of mining products is too high, and there is a lack of market competitiveness and of economic profits."*[36]

35 TAR Specialist Plan, p.474
36 'The search for countermeasures to develop mining in Northern Tibet [*Zangbei*]', Tibet Daily, 4 December 1996

The authors of the TAR Specialist Plan state that in preparing the Plan, data on various minerals had been deleted from its reserves tables and re-classified as 'natural resources' in order to provide a more accurate representation of the TAR's mineral reserves. However, one of the main problems in assessing official information on the mineral wealth of Tibetan areas (from this and other Chinese sources) has been a lack of clarity in terminology. The meaning of a 'reserve' and a 'resource' is often vague; where it is defined it does not accord with western mining classification systems. The traditional reserves system used in the PRC, which was adopted from the former Soviet Union, stressed the geological existence of reserves,[37] as is apparent in the TAR Specialist Plan. In other words reported 'reserves' were geologically proven, but not necessarily economical to exploit. According to definitions used by western mining nations and in international classification systems (for example the UN Framework Classification and definitions used by the Institute of Mining and Metallurgy[38]), resources only become reserves once they have been found to be economically feasible to exploit as well as geologically sound.[39] Increasingly, environmental, social and political considerations are also being taken into account in determining the feasibility of exploiting deposits.[40] Added to the unreliability of geological data are economic and other considerations which mean that many of Tibet's rich mineral 'reserves', as reported by China, may not be considered feasible to exploit by these international standards. Data provided by reports like the Specialist Plan need to be viewed with a great deal of caution.

Over recent years efforts have been made globally to increase international co-operation in standardizing definitions for mineral resources and reserves.[41] In June 1999, the Chinese government decided to reform its mineral reserves administration system according to the United Nations Framework Classification (UNFC) that was proposed by the EU Commission in 1997.[42] Mineral resources are to be classified into three categories: reserves, reserve base, and resources; these are then further divided into 16 subcategories.[43]

37 Pui-Kwan Tse, 'The Mineral Industry of China, 1999', US Geological Survey (USGS) Minerals Information, 2000. The full text of this and other annual reports 1994-2001 can be viewed on the USGS website www.usgs.gov

38 According to its website, the UK-based IMM, founded in 1892, is a professional/learned body for engineers in the minerals industry (www.imm.org.uk).

39 Although the economic feasibility of minerals comes and goes, depending on external factors, for example the market price of minerals.

40 Bill McKay and Ian Lambert, AGSO (Geoscience Australia) and Norman Miskelly, JORC (Australasian Joint Ore Reserves Committee) and CMMI (Mineral Reserves International Reporting Committee), 'International Harmonisation of Classification and Reporting of Mineral Resource', published on the JORC website www.jorc.org (undated).

41 Ibid. According to this report an agreement was made by Australia, Canada, South Africa, the UK and the US, through CMMI, to implement international standard definitions for mineral resources and mineral reserves in 1999. Interest in harmonisation of classification has grown with the emergence of former Eastern Bloc countries.

Another recent reform has been a restructuring of the management of data. Until July 2002, geological data in the PRC were controlled by a number of different organs. However, revised managerial regulations on geological data were due to come into force on 1 July 2002, stipulating that all geological data are to be managed by the Ministry of Land and Resources.[44] This should allow for a more comprehensive databank and greater central control. As part of China's efforts to attract more foreign capital and technology into the mining industry, the regulations will also allow foreign investors to have the same access to geological data as domestic mining companies.[45] According to Xinhua, "*limited access to even basic geological information has been a notable obstruction to the actual flow of foreign investment into the [mining] sector*".[46]

Small gold-mining machine in river-bed, Dawu (Ch: Daofu) county, Kardze TAP, Sichuan © TOTAR, 1997

42 Pui-Kwan Tse, 2000, op cit. The UNFC is designed with a view to bringing closer together all existing classification systems (McKay, Lambert and Miskelly, op cit).
43 Pui-Kwan Tse, 2000, op cit.
44 'Mining data open to foreigners', Xinhua, 4 May 2002. Translated by BBC Monitoring.
45 Ibid. The article states that the public will have access to geological data, excluding data 'concerning the safety of the country'. The data will be classified as non-commercial (free access) and commercial (access subject to payment of a fee).
46 Ibid.

Technology and skilled personnel

> *"The development and utilisation of the mining industry has been severely restricted due to the harsh natural environment, a serious shortage of geological and technical personnel, the seriousness backwardness of science and technology, the extremely low level of prospecting work, the lack of prospecting data and the actual lack of clarity about mineral resources".* General manager of Nagchu Prefecture Mineral Development company[47]

Most of the mining enterprises in Tibetan areas, as is the case throughout China, are medium-small scale operations using very low levels of technology. For example, the TAR Specialist Plan states that:

> *"Tibet has many types of small-scale [enterprises] which use local methods to carry out mining. There are only some individual mines (for example, Norbusa ferrochrome mine in Chusum county and Tumengela coal mine in Amdo county) which use standard mining designs and plans, and which are carrying out mechanisation or partial mechanisation, united with local methods, to mine the surface and underground".*[48]

Mining enterprises in Tibetan areas need to modernise if they are to improve their efficiency and competitiveness. Capital investment is clearly needed in order to improve levels of science and technology, for example to introduce mechanised methods of production. However, according to official reports on the mining industry, there is also a shortage of expertise (technological, managerial and financial), which in itself limits investment opportunities. In order to modernise and develop the mining industry according to current policy new suitably qualified human resources are needed to work in, invest in and run mining enterprises in Tibetan areas.[49]

The problem of what one Chinese academic described as the "*backwardness of education and population quality*"[50] in Tibetan and other western regions of China is not only seen as relevant to the mining industry, but to development in general.

47 'The search for countermeasures to develop mining in northern Tibet [*Zangbei*]', published in Tibet Daily, 4 December 1996
48 TAR Specialist Plan, p.465
49 This section looks at official policy developments that seek to address this issue from the perspective of the Chinese authorities. For a discussion of population and immigration issues related to mining see Chapter 3. For more information on Tibetans' access to employment and training see Chapter 4.
50 Liu Jiaqing, Professor of Demography at the Demographic Research Institute, South West University of Finance and Economics, 'Analysis of the Demographic Environment in the Context of Development of Western China', Beijing Population Research, People's University, 1 July 2000

Improvement of 'population quality' has been one of the main focuses of the Western Development Campaign. In 2001, a 'Ten-Year Plan' for the development of trained personnel in the western regions was formulated.[51] In December 2001, the State Council issued 'Implementation Opinions Concerning Policies and Measures Pertaining to the Development of the Western Region' (referred to below as the 'Implementation Opinions').[52] These opinions outlined the ways in which this development programme will be carried out basically with a dual focus: encouraging people to move into the western areas, while at the same time working to train people in the west and to keep them there.[53]

Since the launch of the Western Development campaign in 1999, the authorities have taken various steps to encourage the 'westward movement of qualified personnel and intellectual resources'. Perhaps most significant is the policy of freedom of movement for both 'qualified personnel' and investors. New regulations allow for people from other parts of the country to move to the west while keeping their original residency.[54] For example, an entrepreneur from Fujian province could now move to Qinghai to set up an aluminium products processing plant, while keeping his residency registered in Fujian province. Alternatively, such people can take residency in the west, but will then be allowed to swap it back to their original home (eg Fujian).[55] These measures will be encouraging to Chinese from the interior and coastal regions who would not necessarily want, by investing in a project in remote Qinghai, to have to register themselves and their families as being resident there. Significantly, the 'Implementation Opinions' also stipulate that if a person has 'permanent legal domicile and a stable job or other reliable sources of income' they can apply for a local urban residence permit. This will enable rural to urban migrants to change their residency status and it may also provide an opportunity for many of the 'floating population' of Chinese migrants, who have no fixed residence registration, to obtain legal status.

51 'Report on the Implementation of the 2001 Plan for National Economic and Social Development and on the Draft 2002 Plan for National Economic and Social Development', delivered by Zeng Peiyan, Minister in charge of the State Development Planning Commission, at the fifth session of the Ninth NPC, 6 March 2002. Full text published in English in the People's Daily, 18 March 2002

52 Published by Xinhua in Chinese, 20 December 2001. Translated and published in English by FBIS, 15 January 2002

53 The 'Implementation Opinions' provide a detailed and comprehensive set of measures, issued at state level. However, many of the measures are clarifications or further developments of previous policy. For example, the 1996 TAR Decisions concerning speeding up development in mining stipulated that the region should send personnel with knowledge and a certain degree of production experience to engage in vocational training and study in education institutes and within enterprises; should attract specialist university and middle school graduates into the industry; should attract professional management experience and production technology; and should recruit and attract highly qualified personnel from elsewhere. ('Decisions of the Tibet Autonomous Region People's Government concerning speeding up of development in mining', 26 October 1996, published in Tibet Daily, 13 December 1996. Referred to below as the 'TAR Decisions').

54 'Implementation opinions', op cit.

55 'Implementation Opinions', op cit. This is in accordance with policies introduced by the Ministry of Public Security in summer 2000, which reportedly stated that all 'investors and professionals working in west China can be registered where they work, and if they wish to return to where they came from they can have their new residence registration go with them.' ('Population, Migration and Birth Control', TIN News Review: Reports from Tibet 2000, TIN, February 2001)

The authorities are also introducing financial and other incentives not only to make working in the western regions attractive to immigrants from the east, but also in an attempt to prevent an outflow of talent from the west. For example, those qualified personnel who move to the western regions to assist in key projects as part of 'counterpart aid'[56] will retain the benefits they received in their home province and may be entitled to an additional subsidy from the host western government. In 2001, a subsidy system was set up by the Ministry of Finance to raise the wage levels for personnel in government organs and institutions in remote border areas and 'areas where conditions are harsh', so that wages in these areas will match or even exceed the national average.[57] Another idea is to set up 'entrepreneurial parks' to attract, in particular, Chinese returning from overseas study.[58]

The authorities have also stated that they intend to develop the 'quality' of the western regions' existing human resources. The 'Implementation Opinions' sets out measures to achieve this, including: making the best use of existing qualified personnel, for example allowing for employment to be extended after normal retirement age; providing subsidies for experts (probably as a means of keeping skilled personnel in the west, rather than losing them to better paid jobs in the east); funding research; and improving living and working conditions (although it is not clear who is expected to bear the financial burden for this). Other strategies include intensification of training, both locally and by sending people either to the eastern region or overseas. The authorities plan to finance this by using 'special state funds' and also by attracting international aid.

The Opinions also say that the state will give 'support' to the development of secondary and college level vocational education, although this is to be funded through commercial bank loans or foreign aid.[59] There is at least one vocational school that has taught classes on mine surveying to Tibetan students.[60]

56 'Counterpart Aid' refers to the matching up of provinces and municipalities directly under central administration in the eastern regions with areas in the western regions. The eastern provinces are obliged to provide aid and assistance to their western partner, for example in building schools, clinics and in the sector under discussion here, in developing qualified personnel and providing assistance for key projects. ('Implementation Opinions', op cit.)

57 'Implementation Opinions', op cit. Since 1999, the state has been increasing the wages and salaries of employees in government departments and institutions. In the Finance Minister's Budget Report for 2001/2002 he made reference to the problems that some county and township governments in the central and western regions were having in paying wages on time. He states that "*To help county and township governments in underdeveloped areas of the central and western regions to overcome difficulties the central budget has provided most of the funding needed to implement the state policy on readjustment of income distribution in these areas since 1999*" (Zeng Peiyan, 2002, op cit).

58 'Implementation Opinions', op cit.

Large gold-mining machine being assembled in Spring, Damangduo gold mine, Dege county, Kardze TAP, Sichuan **© TOTAR, 1997**

59 For example, in January 2002, the United Nations Development Programme (UNDP) agreed to invest US$1 million over three years (2002-04) to train thousands of senior professionals for the development of the western regions, although it is unclear so far what percentage of the trainees will be from the 'minority nationalities'. According to the China Daily, Zhang Xuezhong, minister of personnel said that "*Talented people and specialised technical personnel are crucial factors for the success of the surging and far-reaching exploitation of West China.*" Liang Chao, 'China, UNDP Reach Agreement on Training Professional For Western Region', China Daily 16 January 2002

60 For more information on this course and for the training opportunities available to Tibetans see the section on education and skills in Chapter 4.

"Heaven is high and the emperor is far"[61]: management and control of resources

Management and control over resource exploitation has remained very weak. At every level, government departments and organisations, collectives, State Owned Enterprises (SOEs), the military, the police, herdsmen and farmers have been encouraged to become involved in mining Tibet's rich natural resources (owned by the state) as a way of developing the economy. However, the potential of mineral deposits has not been maximised due to extensive rather than intensive mining practices, thus decreasing the net value of state resources. In addition, the central government has lost a significant amount of revenue to illegal mining and sales, particularly of gold.[62]

According to China's Ministry of Land and Resources, the system for management of mineral resources that developed throughout China has 'obvious defects', providing insufficient control over resource use and allowing for serious malpractice:

> *"[The management system] had the malpractice of different management from different departments usually with conflicts, local protectionism and department division. Furthermore, it got more and more difficult for the central government to perform its duties in such fields as protection of the state's ownership, rational planning and full use of mineral resources."*[63]

Problems of jurisdiction, local protectionism and poor accountability have been very much apparent in Tibetan areas, where local officials, as throughout China, are rewarded for rapid economic growth and social and political stability. As a result, local development has been closely tied in with the personal interests of officials seeking to protect their political positions and gain promotion and/or enrich themselves.

61 An article in Qinghai Daily, reporting on the influx of gold prospectors during the 1980s and early 1990s likened the situation to "*an independent society where 'heaven was high and the emperor far'*". ('Searching the 'Corridor of Gold", Qinghai Daily, 18 January 1999)
62 The long-term environmental consequences of poor management and control, which are very serious, will be discussed in Chapter 5
63 'China's Legal Systems for Management of Mineral Resources', MLR website (undated), www.mlr.gov.cn

Central policy has dictated that mining should be a pillar industry of local as well as regional development in Tibetan areas. For example, in 1996 the TAR issued a set of decisions on accelerating the development of mining, which state that the vigorous development of local mining from prefecture to township level should be one of the major policies for enriching both the people and the county authorities.[64]

As well as the development of larger mining SOEs, for example the Mangya Asbestos Mine in Tsonub, Qinghai province or the Norbusa Chromite mine in Lhoka, TAR, officials at every level have been encouraged to set up new mining enterprises. This has included the establishment of county-level government mines, collective township enterprises, and also the leasing of land for private exploitation as a means to generate revenue. Since the early 1990s in particular, the focus on fast-track economic growth has led to an attitude of development at all costs. The emphasis on making immediate profits from mining, rather than on the solid long-term growth of the industry, does not encourage rational and sustainable development of resources.

Such a high level of official involvement in economic development at all levels has meant that officials see themselves as entrepreneurs as well as government administrators.[65] Local officials are responsible for both economic development and the management of resources and there has been an unspoken deal between levels in the government hierarchy to allow the level below enough flexibility to achieve economic growth.[66] Therefore there is a very low level of accountability within the administrative structure, inevitably leading to irrational exploitation of resources, as well as poor health and safety and environmental standards.[67]

Attempts to crack down on illegal mining have been undermined by the fact that local governments are often benefiting. In some areas at least, illegal and semi-legal mining have constituted a significant proportion of local government income. For example, the local authorities in Chumarleb (Ch: Qumalai) county, Yushu TAP, Qinghai, relied upon income from the thousands of illegal prospectors who flooded to the region during the 1980s and 1990s in search of gold.[68] Officials have also personally benefited from the mining industry through bribes – cash payments or gifts.[69]

64 'TAR Decisions', 1996, see note 53.
65 Kenneth Lieberthal, "*China's governing system and its impact on environmental policy implementation*", China Environment Series Issue 1, published by the Environmental Change and Security Project of the Woodrow Wilson Centre on www.ecsp.si.edu, fall 1997
66 Ibid.
67 See Chapters 4 and 5.
68 See Chapter 5.
69 See Chapters 3, 4 and 5.

As a result, at the same time as reforming and opening up the mining industry, the state is trying to increase its macro-control over the use of mineral resources. Recognising that local protectionism and corruption are threatening China's long-term economic development, official policy in recent years has constantly stressed the need for government administration to be separated from enterprise management. The military has also been banned from engaging in business. However, implementation of this policy will involve a complete overhaul of current administrative structures and directly conflicts with the interests of key officials at all levels.

On 26 November 2001, China's minister for land and resources, Tian Fengshan, announced that fresh attempts were to be made to reassert central control over the mining industry. An acknowledgement was made that "*excessive mining in some regions has caused a serious waste of resources and environmental damage*" and that "*such problems have become a bottleneck for the development of the mining industry and greatly affect local stability and the lives and property of local people.*" The minister blamed "*local protectionism and a lack of control*" for these problems and said that mines that do not hold licences, have the potential to cause environmental damage and waste resources and that operate using 'phased-out technologies' will be closed down.[70] In February 2002, the Qinghai provincial authorities banned all alluvial gold mining in the province, with the exception of three state gold mines.[71]

In international terms most of the mining industry consists of small to medium-scale enterprises, and is inefficient. As with other industrial sectors in China, the authorities are trying to improve the efficiency of the mining sector by restructuring the industry and encouraging smaller operations to form large mining enterprises or 'Enterprise Conglomerates [*qiye jituan*]' in order to pool resources and push for more efficient production.[72]

70 China Daily, 27 November 2001. It should be noted that similar statements have been made in the past, but implementation has been limited. See for example Chapter 4, p. 133

71 The extent to which this ban will be enforced remains to be seen. See chapter 5, pp. 171-179

72 See for example 'Promoting trains of thought and counter-measures for our region's SOE's', Tibet Daily 12 December 2001

'High grading'

As outlined above, the short-term outlook of most mining enterprises and individuals in Tibetan areas has generally led to extensive rather than intensive mining practices. In the drive to make immediate profits, miners have extracted the high grade ores and then moved on to another spot leaving low grade ore behind, a practice known as 'high grading'. The long-term economic and environmental consequences of this kind of mining are very high.

When a mineral deposit is mined for high grade ore and deserted, the low grade ore that remains becomes uneconomic to extract and process. Mineral resources are not renewable and as a result this kind of wastage can have serious implications for the future sustainability of mineral development. Under more rational exploitation plans, a well-run mine would ideally balance costs and profits to maximise the potential of a deposit by factoring in the processing of low grade ore during the life of the mine. China is already concerned over the scarcity of key mineral resources within its territories, including gold and copper, and as a result sees such wastage as a serious problem. The state owns all mineral resources, but is not able to realise the potential benefits of these resources under current mining practices. The TAR Specialist Plan specifically mentions practice of high grading as 'very serious' and one of the key issues that needs to be tackled in developing the mining industry.

There have been several reports on the high grading practices of artisanal and small-scale gold miners.[73] Gold mines are often enriched at the surface, creating the temptation for miners to take the surface ore and then move on to another spot. However, local government-operated and state-owned enterprises have also been responsible for considerable wastage. The Specialist Plan, referring to such enterprises, states that:

> *"The enterprises [producing minerals] pursue profit, extract rich resources and cast aside poorer resources, extract what is easy and cast aside what is more difficult and extract from large [deposits] and cast aside smaller [deposits]".*[74]

73 See Chapter 5
74 TAR Specialist Plan, p.474

The Specialist Plan gives as an example the current practice of high grading in the TAR's chromite and salt mining enterprises. The minimum industrial requirement for high grade chromite ore is equal to or more than 32% chromium oxide (Cr_2O_3) content. However, the TAR's chromite mines produce ore equal to or above 40%. Ore of the grade 32% – 40% is abandoned. Similarly, while minimum industry requirements for boron are 2% boric oxide (Bo_2O_3) content, the TAR's mines are producing ore that is 28% and higher, abandoning ore with a grade of 2% – 28%. The Specialist Plan defines the three key reasons for this malpractice as inadequate management of resources, the high cost of mineral exploitation in the TAR and the fact that "*enterprises are one-sidedly pursuing profits for their own work units [danwei]*".[75]

At the 1998 International Symposium on the Qinghai Tibet Plateau two Chinese scientists from the Commission for the Integrated Survey of National Resources, Song Xinyu and Yao Jianhua, gave a paper on resource exploitation in the Tsaidam basin. They also mentioned the serious issue of high grading:

> *"Existing laws regulating mineral resources extraction are not implemented, so there is no planned exploitation leading to extensive rather than intensive mining. This kind of extraction is focused on the quick extraction of the most readily accessed and highly concentrated portion of the deposit, often rendering the rest uneconomic, even despoiled."*[76]

This problem also extends to the waste of associated mineral resources in a given deposit. For example, gold deposits often have associated deposits of antimony, silver, lead and zinc. It is often the easily extractable gold that can provide the financing for the more costly and long-term extraction of other minerals. However, unless regulated, gold may be extracted and other minerals left behind.

75 Ibid.
76 Song Xinyu and Yao Jianhua, 'Resources Exploitation and Conservation of Qaidam Basin under the Principle of Sustainable Development', a paper presented to the International Symposium on the Qinghai Tibet Plateau, Xining, 24 July 1998

Development and reform of China's gold industry

Chinese central government concern over the generally poor control and management of gold resources has been increasingly evident over the last decade. Gold mining has been a relatively attractive option for local authorities in many Tibetan areas seeking to increase local revenue through mining. Gold is relatively cheap and easy to extract, particularly where infrastructure is poor, there is a lack of finances and technology is outdated. Exploitation throughout China has been widespread and unregulated. However, gold is considered by the state to be of strategic value, and the state loss as a result of irrational exploitation, artisanal mining and illegal sales of gold has been significant.[77]

Even before the founding of the PRC in 1949, the Communists operated a system of monopolistic purchase and management (MPM) for gold and silver in the areas under their control.[78] This in effect meant that they asserted complete control over the purchase, sale, distribution and use of gold and silver in order to protect the new currency of China, the Renminbi (RMB). Once the RMB was established, the authorities retained the MPM policy as a result of China's shortage of foreign exchange reserves. During the 1950s, gold accounted for an average of 60.5 per cent of China's foreign exchange reserves.[79]

Control was tightened in the 1980s due to a continued reliance on gold as foreign exchange during 'development and opening up'. At the same time, China had to reopen the gold jewellery market in order to deal with inflationary pressures.[80] This reintroduced the value of gold as a commodity, not just a product. Starting in the 1990s, the demand for gold jewellery in China started to increase dramatically.[81] As a result producers were looking to maximise profits in line with the entrepreneurial spirit called for by Deng Xiaoping's reforms in 1992. The focus was no longer to be simply production, but production for profit.

77 This section looks specifically at the concerns of the state in protecting its interests and the macro measures that China has taken to reform and develop the gold industry. As such it also provides the background context for further discussion of gold mining and its impact on Tibetan areas in the following chapters of this report. Problems of jurisdiction, accountability and control are discussed as they relate to the Tibetan land and people in the context of ownership and land rights issues (Chapter 3), labour issues (Chapter 4) and environmental issues (Chapter 5).
78 'Administration System and Reform of China's Gold and Silver', China GoldNet www.gold.org.cn, July 2000. The website China GoldNet was established by the Beijing Gold Economic Centre and China Goldnet.co
79 Ibid.
80 Ibid.
81 Ibid.

Most mining enterprises in China had to sell a quota of their output to the state at fixed prices and could then sell any above-quota mining output at market prices. However, the MPM policy meant that gold producers had to sell all their gold to the state via the People's Bank of China (PBC) at a fixed price. This price was set much lower than the international market rate. As a result, gold producers traded unofficially, selling their produce illegally at market price levels rather than the lower fixed state level. However, in China it was not simply a case of a thriving black market. Local governments approved the establishment of local gold exchanges (unauthorised by the PBC) and as such a 'non-governmental' free market developed outside the control of the centre, but with the complicity of local authorities.[82] The price of gold in these non-government exchanges corresponded to international gold prices. As a result of this unofficial trading, by 1993 the Chinese state is estimated by official sources to have been losing 40 per cent of the country's annual total gold output.[83]

In 1993, the central government became aware of what was happening in the domestic gold market and attempted to regain control over the gold industry while at the same time moving towards deregulation of pricing. There was an official crackdown on illegal mining, although it appeared to have limited affect, at least in Tibetan areas.[84] More significant was a reform of the fixed price system so that the PBC could alter China's purchasing price to bring it more closely into line with the international market (a floating price system). The aim was to cancel out the profitability of non-governmental exchanges. There was also talk of deregulation of the industry and the establishment of a Chinese gold exchange, although nothing concrete was achieved at this point. The producers were happy – the price of gold had doubled – and the urgency of reform had faded.[85]

However, in the mid-1990s the international gold price began to fall. Now that the PBC fixed gold price was in line with international prices this had a major impact on the price of gold in China and profits in the industry started to drop. In January and July 1997 the PBC lowered its purchasing price per gram of gold.[86] As a result China's gold producers lost more than US$90.4 million in expected revenues and about one-third of China's 1,200 gold enterprises were losing money.[87]

82 Liu Shanen, 'The progress in the reform for gold marketing in China', Beijing Research Centre for the Development of [the] Gold Economy, September 2000, speech published on China GoldNet, op cit.
83 'Administration System...', China GoldNet, July 2000, op cit.
84 See Chapter 5, pp. 171-179
85 Liu Shanen, 2000, op cit.
86 China Gold, 1998, cited in Pui-Kwan Tse, 'The Mineral Industry of China, 1997', USGS, 1998
87 China Geology and Mineral Resources News, 1998, cited in Pui-Kwan Tse, 'The Mineral Industry of China, 1997', USGS, 1998

In order to protect domestic producers the government instituted a policy of protectionism, keeping the domestic purchase price artificially high. Many traders smuggled gold bullion into China from Hong Kong to capitalise on the difference between domestic and international prices.[88] As with other SOEs, gold mining enterprises lacked initiative because they were protected by state-fixed prices and had access to low interest loans, subsidies, and tax exemptions. This resulted in a heavy fiscal burden for the government who were supporting inefficient and unprofitable enterprises. The PBC announced on 20 February 1998 that the official purchasing price paid by the PBC to gold mine producers would be reduced by 8.5 per cent (to 80.5 yuan/US$9.7 per gram) – the most significant reduction since China started to liberalise its gold policy in 1993.[89] According to the World Gold Council, this represented a major reduction in state subsidy of the mining industry of the PRC. It was hoped that it would force miners to reform and modernise; one consideration was preparing the industry for imminent World Trade Organisation (WTO) entry when it would have to face international competition.

By 1999 it was clear that the industry needed to be reformed and the decision was made to move towards deregulation and the opening of a gold exchange. Reform measures were introduced into the gold industry such as a relaxation of controls on the purchase of recycled gold so that authorised jewellery firms were able to purchase directly. A silver exchange, viewed as a trial for the gold exchange, was due to be launched in Shanghai on 4 January 2000 following the formal abolishing of the state monopoly over silver in November 1999, but its opening was delayed pending a government decision on value-added tax levels for miners and processors.[90] The PBC reduced gold prices four times in 2000 in line with fluctuations on the world market and by the end of the year the price had fallen to 72.6 yuan (US$8.7) per gram, from 82.8 yuan (US$9.9) at the beginning of the year.[91]

Complete deregulation of the gold industry initially involves the establishment of a gold exchange to gradually replace the state monopoly over gold production, purchasing and allocations. The Shanghai gold exchange opened in November 2001 for trial operations and was expected to be officially launched sometime in 2002,

88 Pui-Kwan Tse, 1998, op cit. According to John Adams, director of China Financial Services and author of the report 'Deregulation of the Gold Market in China', levels of inward flows of gold may have fluctuated between 50 and 400 tonnes in any one year over the past decade. The hidden cost of this to the Chinese balance of payments may have been as much as US$4bn (John Adams, 'Deregulation of the Gold Market in China, Executive Summary', 2000, published on China Online www.chinaonline.com).

89 'Council Regulatory Work Helps Chinese Authorities to Reform Gold Market', World Gold Council news release, 26 March 1998

90 AFP, 17 March 2000

91 China Daily, 19 December 2000

albeit on a limited basis. The PBC will initially retain its business of purchase and sales, but on the basis of free trade and according to market prices. The second stage will be to open up the domestic market, which means that the central bank will no longer purchase gold directly from the producers and all gold produced will be traded on the gold exchange. The third step is to integrate the domestic market into the international market, which will require the full convertibility of the RMB. Once this is achieved the central bank will no longer have the role of adjusting prices on the domestic market – prices will become controlled by the international market – and China will release its control over the import and export of gold.[92]

China officially acceded to the WTO on 11 December 2001. Official reports on the impact of WTO entry to the China's gold industry outline two key challenges. The first is that while China has an obligation under WTO rules to open its gold market to the outside world, it has yet to establish and consolidate its domestic market. The other challenge, affecting gold producers and enterprises more directly is the increased competition they will face from gold firms in developed countries with higher levels of technology, more capital and lower production costs. The opportunities brought by WTO entry will be greater access to foreign capital and technology through the abolishment of restrictions on foreign involvement in gold mining and production in China, and the equalising of treatment of foreign and domestic investors and enterprises.

As with other SOEs, alongside the opening of the gold market, China is attempting to reorganise and reform the management structure of gold mining enterprises in order to increase efficiency, weed out weaker enterprises and increase the ability to compete internationally. According to the China Business Weekly, China is going to merge its 1,200 gold mines into 12 'internationally competitive groups' by 2005.[93] This is similar to previous attempts to consolidate other sectors of industry in China. Enterprises are also being encouraged to explore possibilities of developing gold resources abroad.[94]

This will effectively move the industry towards larger-scale exploitation, leaving smaller local enterprises unable to compete. In 1998, President of China National Gold Corporation, Cui Xiwu, acknowledged that the reform of the gold industry would "*inevitably lead to the closing down of many small gold mines*". As with reform in other sectors of China's industry the aim is to weed out weaker enterprises that are costing rather than making the state money. Cui made this quite clear:

92 There have been various reports published on the deregulation of the gold industry in China. For example, see John Adams, 'Deregulation of the Gold Market in China', 2000. See also Liu Shijin, 'Opening China's Gold Market in a New Era – Related Policy Research and Suggestion', the Industrial Economics Research Department, Development Research Centre (DRC) under the State Council of the PRC, October 2000. The report was commissioned by the World Gold Council and the full text is available on its website www.gold.org

93 Report carried by AFX-ASIA 10 December 2001, on China GoldNet, www.gold.org.cn

94 The same is true for other sectors of the minerals industry. For example, in July 2002, Baoshan Iron and Steel Group Corp signed a US$670m deal with Australian company Hamersley Iron Pty Ltd to mine iron ore in Western Australia ('Chinese, Australian firms sign iron ore mining deal', Xinhua, 1 July 2002).

> *"In the short run, only those with real strength and capabilities will survive and thrive. I welcome this as crystallising the spirit of a free market and I am sure it will induce healthy development for the gold market in the long term."*[95]

The impact of these reforms on gold mining enterprises in Tibetan areas is likely to be significant, as the majority of them are unlikely to be able to compete. The gold mining industry in Tibetan areas is mainly comprised of small-scale enterprises, and at least in some areas has remained largely outside the control of central or regional/provincial authorities. The TAR, for example, has only one known alluvial gold deposit that could be commercially feasible to exploit, located in Nagchu prefecture.[96] Qinghai province appears to have led the way in attempts to restructure the gold industry and close down small mines. Alluvial gold mining was banned throughout the province in February 2002, with the exception of three key mines. While the environmental benefits of tighter control over gold mining are unquestionable, the restructuring of the industry, if successful, will mean the closure of many of Tibet's collective and local government mines, or their take-over by larger enterprises. This could effect local government income in the same way that the prohibition on deforestation at the end of the 1990s had a serious impact on some local economies, particularly in western Sichuan.

Gold miners in riverbed, Nyarong **© ART/TOTAR**

95 World Gold Council news release, 26 March 1998, op cit.
96 See Chapter 1, p. 25

Gold-mining settlement on the Tibetan grasslands © TIN, 2000

Tibetan villagers gathered to sieve for gold, north of Minyag (Ch: Xinduqiao), Dartsedo (Ch: Kangding) county, Kardze TAP, Sichuan © TOTAR, 1997

Investment

"Tibet's economy is backward and lacks financial strength. It is difficult to accumulate more funds in order to develop the mining industry – an industry in which the turnover of funds takes a comparatively long time. At the same time, the technical equipment of all mining enterprises is old, the workforce has a low cultural level, there is a lack of all types of engineering and technological personnel, including management personnel and the investment climate is deficient. As a result, economic benefits in these enterprises are low, and so there is no capability of raising more funds to extend production." TAR Specialist Plan[97]

The attraction of new sources of capital and technological investment is vital if the state is to succeed in developing Tibet's mining industry in line with official plans. During the early development of the mining industry in Tibetan areas in the 1950s and 1960s, projects were centrally funded and state-run and largely depended on large quantities of community/compulsory and prison labour. As with the rest of industry, the enterprises were production focused, rather than profit-orientated. Due to economic reforms introduced since the early 1980s and marketisation of the Chinese economy, the state has been seeking alternative sources of funding for mining.

According to the Mineral Resource Law that came into force on 1 January 1987, state-owned enterprises would remain the 'principal force' in developing the mining industry, but collective mining enterprises would also be encouraged. Individuals were also permitted to mine under supervision.[98] Local governments were increasingly encouraged to invest in medium to small-scale mining ventures as a way of developing local economies. Following Deng Xiaoping's 'Spring Tide' reforms of 1992, local governments and existing mining enterprises found themselves under increasing pressure to operate on a profit-orientated basis, as with other industries. China's mining industry started looking for private investment, both domestic and foreign, as a way of obtaining the necessary funds for development. By the end of the 1990s, the Mineral Resources Law had been amended to allow for private domestic and foreign investment and the TAR, for example, was seeking investment for its major mining projects both internally from domestic mining companies and also from foreign companies.

97 Op cit, note 1

98 Mineral Resources Law of the People's Republic of China, adopted at the Fourth Session of the Sixth National People's Congress, 12 April, 1986, effective as of 1 January 1987.

Article 4 The state-operated mining enterprises shall be the principal force in exploiting mineral resources. The state shall guarantee the consolidation and expansion of state-operated mining enterprises.

The state shall encourage, direct and help the collective mining enterprises of township and towns to develop.

By administrative measures, the state shall direct, help and supervise individuals to conduct mining according to law.

However, at the same time that China has been opening up to the idea of private enterprise investment, both domestic and foreign, the central authorities have remained reluctant to remove restrictions on exploration and mining. The interests of the state have become, to some extent, divided, and early reforms in the 1990s were half-hearted, while local initiatives to attract funding have been squashed at a central level due to an unwillingness to offer much to potential investors. The investment environment in China does now appear to improving as a result of reforms over the past decade, although foreign mining companies are still likely to approach investment in China, and particularly in remote western regions like Tibet, with a certain degree of caution.

Although there have been considerable leglislative and regulatory reforms over the last ten years, it will remain difficult to attract investment into the mining industry in Tibetan areas. The problems outlined above – remoteness, poor infrastructure, low levels of science, technology and data and poor management – are all major obstacles to attracting commercial investment.

One of the most serious problems that China faces in attracting foreign investment, particularly relevant to such remote areas as the TAR, is the increasing competitiveness of the international minerals market. According to the manager of the World Bank Mining Department, Peter van der Veen, the trend in average real metal prices has shown a continued decline over the past 35 years.[99] Chinese economists Hu Angang and Wen Jun have also highlighted the global drop in prices for energy and raw materials, which they argue will result in "*a surplus of productive forces, a sharp fall in demand and [...] a broad range of price falls in international markets*" for mineral products.[100] International competition has also increased as a result of emerging minerals industries in South America, Eastern Europe and Sub-Saharan Africa, and also due to technological changes that have made low-cost production increasingly possible where new technology is applied.[101]

99 Peter van der Veen, 'Attracting Private Mining Investment', paper given at the Round Table Conference on Foreign Investment and Mining Development in Western China, Xinjiang Uighur Autonomous Region (XUAR), 13-14 October 2000. Papers from the conference are published on the MLR website, www.mlr.gov.cn. According to a UK-based risk analyst, over the last few years the major mining companies have not been looking to expand their gold mining operations. Due to the fall in price of gold, many safe, guaranteed, reserves have been left unexploited, or operations have been closed down, many of these in South Africa. If the price of gold rises, these will be re-opened and exploited. This means that there is stiff competition making it all the more difficult for China to attract foreign investment from major mining companies.

100 Hu Angang and Wen Jun, 'The Problem of Selecting the Correct Path for Tibetan Modernisation', China Tibetology, 2001 Vol. 1, pp.3-26

101 Sak Kupasrimonkol (Associate Director, Credit Review Department, International Finance Corporation, World Bank Group, Washington DC), untitled paper given at the Round Table, XUAR, October 2000, op cit.

Mining and human rights[102]

"In the past decade, the mining and minerals industry, like other parts of the corporate world, has come under tremendous political pressure ...to improve its social, economic, and environmental performance and its transparency." – Mining, Minerals and Sustainable Development (MMSD) Project[103]

The environment in which mining companies operate has changed over the past decade and there is now an expectation from civil society, particularly in the West, that companies take on board a vast range of human rights issues in their enterprises in other countries. Even if the economic environment is made attractive enough to encourage foreign investment, there are also political and social aspects to mining in Tibet that may increase the risk for foreign companies, particularly major mining companies.

Around the world, benchmarks of best practice in mining are often expressed as the necessity of miners to achieve and report on a 'triple bottom line' of not only profitability, but also environmental outcomes and human impacts on effected communities. While the traditional view among mining companies has been that it is the host government's responsibility to safeguard the rights and interests of local populations, this is changing to some extent in the face of public pressure. Transparency, accountability and participatory development have all become key concepts in the 'sustainable development' debate in the minerals industry. The aspirational ideal is to create an industry that will benefit local communities, limit damage to the environment, but remain commercially feasible.

While there may be limited vocal opposition to a mining project from local communities in Tibet due to the firm measures of control employed by the Chinese authorities, foreign mining companies operating in Tibetan regions are likely to face challenges from NGOs seeking to represent the interests of Tibetans. For example, the London-based Free Tibet Campaign targeted BP following the company's decision to buy shares in PetroChina, the Chinese petroleum company responsible for constructing a gas pipeline from Sebei in the Tsaidam basin to Xining and Lanzhou. Pressure has also been put on Australian junior Sino Mining International (SMI) by Australian Tibet supporters as result of SMI's interests in gold resources in Tibet (see p.74).

102 For further discussion see Chapter 6
103 'Mining, Minerals and Sustainable Development Project Draft Report for Comment: Executive Summary', IIED, 4 March 2002

Opening up mining to foreign investment

Although the PRC's legal system is still considered to be in the early stages of development,[104] China has taken considerable steps over the past decade to open the mining sector to foreign investment and in accordance with this to reform and develop relevant legislation. The legislation that has been developed reflects a change in the balance of conflicting state interests (the need for foreign capital and technology versus the desire to retain control over resources) with a gradual move towards easing restrictions on foreign involvement and increasing incentives for investors. It also reflects the central government's shift of focus away from facilitating development in eastern China and towards development of the western regions at the end of the 1990s.

The Chinese government did not begin to develop legislation dealing with foreign investment in mining until 1993. Prior to this most legislation was designed to apply to domestic mining enterprises.[105] The UN-sponsored Round Table Conference on Mining, held in Beijing in 1993, saw the creation of the first draft of regulations to govern foreign investment in mining, and the same year US oil company Exxon (now ExxonMobil) was awarded petroleum exploration rights in the Tarim Basin, Xinjiang. However, these draft rules remained in contradiction to the Mineral Resources Law (1986).[106] While implementation rules for the Mineral Resources Law (MRL) issued in 1994 formally authorised foreign participation it wasn't until 1996 that the MRL was finally revised. According to the revised MRL, which went into effect on 1 January 1997, foreign investors receive the same treatment as domestic investors with regard to exploration rights, mining rights and transfer rights.[107] One of the most important reforms, crucial to attracting private investment, has been the development of legislation governing the transfer of mining rights, prohibited under the 1986 Mineral Resources Law which stated that: "*Mining rights may not be sold, leased, or put in pledge*" (Article 3).

Despite the signal that the mining industry was open to foreign investment in 1993, it soon became clear that Chinese policy remained very restrictive, particularly for those minerals considered to be of strategic value, including gold. Initially, considerable interest was generated amongst foreign mining companies, particularly regarding possibilities for gold exploration and exploitation.[108] However, this enthusiasm was

104 See for example Paul D. McKenzie (Partner, Perkins Coie LLP, Hong Kong) and May K. Leung (LL.B candidate, University of Victoria), 'New Rules Mean New Prospects in China for International Mining Companies', Perkins Coie, March 2001
105 Michael J. Moser (Partner, Baker & Mckenzie), 'Recent developments in mining law in the People's Republic of China', Energy Law – Special Supplement, 1995
106 Ibid. Moser states that provincial governments, including Yunnan and Xinjiang, incorporated many of the provisions found in the draft regulations, but that *in the absence of national legislation the validity of these local rules was open to doubt.*
107 William L. MacBride, Jr. (Partner, Gough, Shanahan, Johnson & Waterman, Montana) and Wang Bei (Division Chief, Mining Rights Transfer, Department of Geological Exploration, MLR, PRC), 'Chinese Mining Law Overview', InfoMine, 2001
108 McKenzie and Leung, 2001, op cit.

dampened somewhat when, in June 1994, the State Council issued Document No. 64, which set out clear limitations for any foreign involvement in gold mining. The Document stated that foreigners were only permitted to invest in the mining of gold reserves 'which are of a low grade' and are 'difficult to work using domestic mining technology'.[109]

Other factors also suggested that China was not yet prepared to fully commit to foreign involvement in the mining industry. Foreign investors were only permitted to operate in China as part of equity or cooperative/contractual joint ventures with Chinese partners. The approval procedures for such mining projects remained very complicated and lengthy, involving different government departments and different levels of authority. Most notably, there was a lack of clarity with regard to exploration, mining and transfer rights. Foreign investors were still not guaranteed sufficient protection of their interests or offered preferential policies to make projects attractive enough to warrant huge investments in such a risky industry. Most projects for which foreign involvement initially looked promising failed to get off the ground.

For example, in 1996, Asia Mineral Corp, a junior Canadian mining company was the first foreign mining company to receive approval from the State Council to form a joint venture to develop a gold mine in China. The joint venture, with Shandong Zhaoyuan City Gold Corp. was issued a business licence in October 1997 to develop the Yingezhuang Mine in Zhaoyuan city, Shandong province. However, Asia Mineral Corp terminated the joint-venture contract in 1998, stating that Zhaoyuan breached numerous principles and provisions of the contract; in addition, the joint-venture was not given the mining licence or preferential tax policies originally offered.[110]

Following the opening up of China's mining industry, the authorities had also initially appeared eager to attract foreign investment to develop the Yulong copper mine in Chamdo, TAR, the only known deposit in the region with world-class potential.[111]The authorities hosted visits from Western companies, including BHP[112] and Cyprus Amax. Although there were obvious problems with developing the deposit due to its remoteness and altitude, the US-based Cyprus Amax[113] still considered the deposit to be large enough to be of potential interest – and the only attractive mine in Tibet. However, Cyprus Amax withdrew their interest in the project due to unwillingness on the part of the authorities to relinquish their own interests in the project. According to a reliable source within the industry, the Chinese government had decided to develop the mine themselves:

109 Moser, op cit.
110 Pui-Kwan Tse, 'The Mineral Industry of China, 1997', USGS, 1998
111 See Chapter 1, pp. 31-33
112 Now BHPBilliton.
113 Cyprus Amax has now been taken over by and incorporated into the Phelps Dodge Corporation.

> *"They like to attract foreign investment but don't seem to take the next step, which is to approve equity and interest in a project. The security of an investment depends on the security of title, which means rights to the mine based on a contractual arrangement."*

However, China continued to reform the legislative and regulatory environment following the revision of the Mineral Resources Law in 1996, by developing legislation governing exploration, mining and transfer rights, setting procedures for foreign investment, and offering preferential policies.[114] The restrictions placed on foreign investment in gold mining as stipulated by Document 64 were also removed.

The most significant step so far towards improving the investment environment for foreign companies came in December 2000, when the State Council issued "*Several Opinions Concerning Further Encouragement of Foreign Investment in Exploration of Mineral Resources other than Oil and Gas*".[115] Of particular note is a provision that eliminates the need for foreign investors to operate as part of a joint venture with Chinese partners; a foreign company can now obtain an exploration permit in its own name and operate on its own. Also significant is a requirement for application procedures to be simplified and speeded up. The Ministry of Land and Resources (MLR) must now respond to applications for exploration and mining rights within 30 days.[116] Earlier in 2000, approval rights for industries encouraged by the government were delegated to the provincial level, with the exception of the exploration and mining of gold and the development and utilisation of rare earths, which remain under the jurisdiction of the state.[117]

114 For example, on 12 February 1998 the State Council issued three sets of regulations: 'Regulations on Registration for Mineral Exploitation'; 'Regulations on Registration for Exploration of Mineral Resources'; and 'Regulations on Transfer of Exploration Rights and Mining Rights'. On 14 April 1998, China issued a new policy document outlining permitted foreign mining entities and the minerals available to foreign investments and providing for a set procedure for foreign investment. (MacBride and Wang, ibid). In December 1999 the Ministry of Finance enacted regulations governing the accounting treatment of the exploration and mining rights of enterprises and geological exploration organisations (Shou Jiahua, 2000, op cit).

115 McKenzie and Leung, 2001, op cit. China's Vice Minister for Land and Resources, Shou Jiahua had already outlined some of these new proposals (approved by the State Council but not then enacted) at the 'Round Table Conference on Foreign Investment and Mining Development in Western China', held in XUAR in October 2000 (Shou Jiahua, October 2000, op cit).

116 The extent to which this will be implemented in practice remains to be seen.

117 Shou Jiahua, op cit. According to Shou, in a trial run of this policy, Yunnan province was given the authority to approve foreign investment in provincial mineral exploration and development in 1999, resulting in the signing of agreements with foreign companies worth a total contractual foreign investment of more than US$30 million. He gave no further details.

The 'Several Opinions' document also provides confirmation of policies that have been formulated in recent years, including the provision that a foreign company discovering a mineral deposit as a result of its exploration work holds a legal priority to mine that deposit.[118] Incentives that have been introduced include the same preferential tax treatment currently available to other foreign enterprises in China, preferential treatment in importing exploration equipment, and the ability to treat exploration expenses as deferred assets.

Development of the western regions

New reforms and legislation introduced since the launch of the Western Development Campaign in June 1999 to further clarify and improve the investment environment in the minerals industry, have included preferential policies to encourage foreign investment in the western regions of China.[119] According to the 'Several Opinions', new tax incentives and preferential policies specifically targeted at guiding foreign investors to western China include exemption from exploration and mining rights usage fees for one year, with a 50 per cent reduction in those fees for the following two years. If the mineral resource being mined is one of those encouraged by the government, foreign investors will be exempt from paying mineral resource compensation fees for five years.[120]

As well as the attempts to attract foreign investment, the central authorities are attempting to increase investment into the mining industry in the western regions by transferring knowledge, expertise and capital from the richer eastern provinces of China. One of the ways in which they are doing this is to set up 'Counterpart Aid' projects. This refers to the matching up of provinces and municipalities directly under central administration in the eastern regions with areas in the western regions. The eastern provinces are obliged to provide aid and assistance to their western partner. This system is used for developing education and healthcare in the western regions,

118 McKenzie and Leung, 2001, op cit. Other provisions that have been confirmed include: foreign companies may count advanced technology or equipment as investments; and they may buy exploration and mining rights legally held by large and medium-sized domestic companies.

119 This has been true of investment in general. For example, in August 1999, the State Council published proposals to further encourage foreign investment in the central and western regions of China. In June 2000, the Chinese government published the Catalogue of Advantageous Industries for Foreign Investment in Central and Western China. All the listed industries belong to the 'Encouraged Class' of industries as listed in the 'Catalogue for Guiding Foreign Investment in Industries' and as such enjoy preferential policies such as exemption from tariff and duties on imports of necessary equipment. Shou Jiahua, October 2000, op cit.

120 Other circumstances under which foreign investors can be exempt from mineral resource compensation fees include the mining of associated deposits and the use of technology to exploit difficult deposits. The 'Several Opinions' also introduce protection for the foreign investor in a joint-venture enterprise in the form of a requirement that exploration and mining rights to be contributed by the Chinese party be evaluated by a licensed institution and be priced fairly. (McKenzie and Leung, 2001. Ibid.)

but is also being used within the mining industry. For example, the Henan Provincial Mining Authority has been teamed up with Dengchen (Ch: Dingqing) county in Chamdo prefecture. The Henan authority carried out a five-year geological survey in the area from 1990-1995 and in 1994 Dengchen county government and the Henan inspection team formed the Dengchen County Mineral Development Company. They have since established a jewellery handicraft factory, two ferrochrome mines (Dongyangti and Xiyangti) and the Zijiapu colliery.[121]

Measures that have been taken to spur the development of the minerals industry in China, particularly in the Western regions over the last few years, also include clearer guidelines and preferential policies for SOEs and private domestic investors. For example, money paid for prospecting rights or mining rights in the western regions following state-funded prospecting may now be converted into the state capital of a state-owned mining enterprise or prospecting unit under certain conditions.[122] A unit that prospects for, mines or extracts mineral resources in the western region may apply to be exempt from the prospecting rights use fee and the mining rights use fee, or to pay the fees at a reduced rate under certain conditions.[123]

Foreign involvement in mining Tibet

China is receiving international support for its efforts to develop the mining industry of Tibet and other 'minority nationality' areas, despite the fact that investment in China's minerals sector remains a high-risk prospect for foreign mining companies – particularly in the remoter western regions. For example, the World Bank has given its support to the Chinese authorities in the drive to develop the western regions through resource exploitation. The Bank, which has a mining department, is committed to promoting private investment in mining in developing countries. It states that it is

121 'Enlightenment promotes openness; openness promotes development: Dingqing county uses mineral development as the spearhead to develop township enterprises and mining', Tibet Daily, 8 January 1997. Names in Chinese as received.
122 'Implementation Opinions', 2001, op cit. One of the following conditions has to be met. 1) The resources being prospected, mined or extracted are petroleum, natural gas, coal-bed gas, iron-rich ores, high-grade manganese ores, ferrochrome ores, copper, nickel, gold, silver, sylvite, platinum metals, or underground water. 2) The minerals are being prospected, mined or extracted in areas designated by the state as key poverty alleviation areas or as key development areas. 3) A large or mid-sized mining enterprise is prospecting for successor resources because of resource depletion. 4) A state-owned mining enterprise has obtained approval to reorganise itself by adopting the shareholding system or by entering into a partnership with a foreign company and the unit that owns the state capital wishes to buy into the enterprise by paying for its own stake with prospecting rights or mining rights. 5) Or it is verified that a state-owned mining enterprise cannot afford to pay for prospecting rights or mining rights because of a natural disaster or other factors beyond its control.
123 'Implementation Opinions'. One of the following conditions has to be met. 1) Same as note 49. 2) Same as point 3 note 49. 3) New technology and new methods will be used to raise the level of comprehensive utilisation (inlcuding exploitation of low-grade ores and residual ores in old mines). 4) Other situations approved by the government department in charge. The enterprise may be exempt from the prospecting and mining rights fees for the first year (and for the mining rights fee also for the capital construction phase), and pay only 50 per cent in the third and fourth year, then 75 per cent from years four through seven. No use fee is payable in the year when the mine is closed.

Silver mining in eastern Tibet **© TIN**

'uniquely placed' to lower the risk for investors in developing countries, particularly junior mining companies and to ensure that mining can act as poverty alleviation, "*through responsible, cutting-edge investment practices in partnership with local communities*".[124] The Bank has co-sponsored conferences in China in 2000 and 2001. The former, held in Xinjiang Uighur Autonomous Region (XUAR), was aimed at improving the investment environment for mining in western China.[125] The second, held in Xi'an in September 2001, was more focused on the environment.[126]

However, the Tibetan areas of the PRC are among the most remote of China's western regions and at present have the weakest infrastructure. These factors, when combined with the other obstacles to development discussed above, mean that, in the near future at least, investment in the exploitation of non-oil and gas minerals in Tibetan areas is likely to remain largely in the hands of the state and domestic private enterprises, with some investment from junior foreign mining companies.

124 See Chapter 6 for a critical analysis of the development model in Tibet, including the premise that resource exploitation in Tibetan areas functions as poverty alleviation. The World Bank has previously found itself facing international criticism regarding its involvement in China's poverty alleviation strategies, notably the transfer of 58,000 poor Hui Muslim and Han Chinese farmers from north-east Qinghai to Dulan county in Tsonub M&TAP. For the most recent TIN report on this project, which is going ahead without Bank funding, see 'Resettlement and urban construction in former World Bank project county', TIN News Update, 14 February 2002.

125 Round Table Conference on Foreign Investment and Mining Development in Western China, XUAR, 13-14 October 2000. Papers given at the conference have been published on the Ministry of Land and Resources website, www.mlr.gov.cn

126 'Improving the Investment Climate and Sustainability of Mining in China', September 20-22, 2001, Xi'an, Shaanxi province. A list of papers given at the conference can be viewed on the World Bank wesbite, www.worldbank.org

According to available information, there are currently no foreign companies directly involved in mining in the TAR. While foreign companies are similarly unknown to have reached the extraction stage of non-oil and gas mining projects in other Tibetan areas, there has been foreign investment in exploration and at least two foreign junior mining companies have entered into joint venture agreements with Chinese partners.[127]

In June 2001, a subsidiary of the Australian based junior mining company Sino Mining International (SMI) signed a contract with Deyang Mineral Resource Development Company to form a joint venture, 'Sino Mining Jinkang Ltd', to explore for gold in the Jinkang area, located in Kakhog (Ch: Hongyuan) county, Ngaba (Ch: Aba) Qiang and Tibetan Autonomous Prefecture (Q&TAP), Sichuan. The site had previously been explored by BHP and rated as one of their top three gold exploration properties.[128]The subsidiary company, Sino Mining Sichuan Pty Ltd, holds a 51 per cent interest in the Jinkang project. This will increase to 75 per cent if they complete an investment of US$14.5 million and have already finished a feasibility study. SMI had previously owned an 87.5 per cent share in a joint venture to explore the Tanjianshan gold deposit in Dachaidan, Tsonub M&TAP, Qinghai. Once in production, Tanjianshan was expected to produce about 100,000 ounces (approx. 2.8 tonnes) of gold per annum.[129] However, following completion of the exploratory drilling programme, SMI announced in February 2001 that it had pulled out of the project because the estimated mineralisation of the site indicated that the project would not be sufficiently profitable to warrant further exploration or development at that time.[130] One of the ore sites in the Tanjianshan deposit, Jinglonggou is now being developing by a domestic company.[131] SMI were also involved in putting together the recent study of gold deposits in the western Qinling belt, 16 of which are located in Tibetan areas in Sichuan and Gansu.[132]

127 In the oil and gas sector, Italian oil company Agip has signed a contract with China National Petroleum Corporation to become operator of the gas-fields in the Sebei block of the Tsaidam basin, Qinghai. The block, which has an area of 6,998 sq km, was being operated by PetroChina. ('Agip drilling on its northwest China Sebei Block', Oil and Gas International, 23 April 2002). British-based multinational BP attracted considerable attention from Tibet campaign groups when it purchased a large share of PetroChina, the company responsible for constructing the Sebei-Xining-Lanzhou gas pipeline. The pipeline was completed by the end of 2001 ('Key Gas Project in Western China Begins Operation', People's Daily 13 December 2001)
128 Nicole Sonter, 'Contract signed by Sino Mining Sichuan Pty Limited', SinoMining press release, June 2001. According to the press release, previous work by BHP (now BHPBilliton) and a local exploration group had outlined a 70km long zone of strong gold anomalism in regional stream and soil sampling from Shuajingsi to Xinkangmao. A consultant geologist told TIN that he believed BHP pulled out because they had experienced problems exploring in China and/or were moving away from gold exploration. A UK-based risk consultant also said that although the company had been very expansionist in the early 1990s, it has been experiencing financial difficulties in the last few years and has been 'drastically cutting costs'.
129 Sino Mining Limited, project information, 2000, on the SMI website www.sinomining.com.au
130 'Sino Mining Ltd has no current plans to develop mine asset in Tibet', Australia Tibet Council Media Release, 5 February 2001
131 Dachaidan Jinlong Mining Development Company. See Chapter 1, p.25
132 See above, pp. 46-47 and Chapter 1, p.27

During the 1990s, a Canadian junior mining company, Breckenridge Resources Ltd., had interests in three silver-base metal deposits in Bathang (Ch: Batang) and Payul (Ch: Baiyu) counties, Kardze (Ch: Ganzi) TAP, Sichuan, although these were never developed. Breckenridge, which is a subsidiary of Canadian junior mining company Athabaska Gold Resources, formed a joint venture with Baiyu Xinyuan Mining Company[133] and Norsang Corporation Ltd.[134] to develop the Xiacun deposit in Payul, in which Breckenridge owned a 68 per cent share. The company had carried out a prefeasibility study in 1996 and had reached an agreement on the terms of the project with its joint venture partners in 1997, following which the joint venture applied to the central government for approval.[135] In April 1997, the company announced that it had also signed a memorandum of understanding on two further silver-base metal deposits, the Xisai deposit in Bathang and the Gayiqiang deposit in Payul.[136] In April 1999, Breckenridge and its joint venture partners received approval from the Chinese Central Government State Planning Commission to develop the Xiacun project.[137] However, in July 2000, Breckenridge announced that it had sold all of its rights, obligations and interests in the Xiacun project to APAC Minerals Inc., another Canadian junior.[138] The reason for Breckenridge's withdrawal is likely to have been linked to financial problems; Breckenridge, and also its parent company, Athabaska Gold, subsequently appear to have folded.[139] APAC Minerals told the Canada Tibet Committee in June 2002 that the Xiacun project had been 'terminated'. Although there is no longer Canadian involvement in Xiacun, there are unconfirmed reports that mining is going ahead at the site.

133 A consortium comprising China Non-Ferrous Metals Chengdu Corporation, Sichuan Provincial Bureau of Geology and Mineral Resources, Kardze Non-Ferrous Metals Industrial Corporation and Payul county Industrial Resources Corporation, each with a 25 per cent interest.

134 A private company incorporated in Hong Kong.

135 'Athbaska Gold Resources Ltd. – Xiacun silver-base metal project Sichuan province, China – of Breckenridge Resources Ltd.', Athabaska press release, 11 Feburary 1997

136 'Athabaska Gold Resources Ltd. announces memorandum of understanding on Chinese projects reported by Breckenridge Resources Ltd.', Athabaska press release, 15 July 1997

137 'MineSCAN 2000-2001', Caratax Management Ltd, published on www.homestead.com

138 Canada Tibet Committee (CTC). APAC Minerals' most noteworthy project is in Argentina, supported by Rio Tinto.

139 Ibid. According to CTC, in January 2001, the Canadian exchange regulatory agency announced that trading of Breckenridge stocks was inactive and that it had until 11 July 2002 to comply with a variety of requirements before its stock could be traded again. The company appears to no longer exist.

Gold-mining camp, Ngari (Ch: Ali) prefecture, TAR, fossil-fuel generator in the foreground © TIN

Gold mining, Dartsedo county, Kardze TAP, Sichuan, spring 1999 © TIN

3. Resource ownership and land use

Mining is one of the most sensitive land issues throughout the world, particularly when territory marked for mining is used by an indigenous population whose local economy, society and religious and cultural beliefs are tied to the land. The immediate impact of resource exploitation on these indigenous populations is often negative and can include the expropriation of land for mining, the in-migration of workers, and damage to the local environment that can affect livelihoods, health and can also transgress religious and cultural beliefs. There is growing international pressure on the mining industry and governments to recognise and respect the rights and interests of indigenous populations, who often remain on the fringes of national and international development.

The majority of Tibetans rely upon the land for their livelihoods and as a result the mining of pasture and arable land has a direct impact on their lives. However, Tibetans living under the administration of the People's Republic of China (PRC) have little if any chance of engaging in decision-making processes over the use of local land and resources. With the Chinese Communist Party's assumption of control over Tibetan territory came an assertion of State ownership of all mineral resources. Today, the interests of the State and/or local officials often eclipse the interests and concerns – practical or religious – of nomads and farmers.

The first section of this chapter outlines the Chinese constitutional and legal framework regarding ownership and rights to mineral resources and land. It establishes that under this framework there are no 'Tibetan' resources and that Tibetans are left with little opportunity to benefit from the exploitation of minerals in their localities, while they remain vulnerable to the State's power to move anyone on to their lands or to force anyone to relocate. The second section focuses on immigration, an issue of great importance and urgency for many Tibetans. It describes the colonisation of the Tsaidam basin during the early years of the PRC, the increasing flow of Chinese migrant labour to the mining industry over the last two decades of economic reforms and the influx of tens of thousands of artisanal miners into Tibetan areas. The third section looks at the expropriation of land from farmers and nomads and the threat this can represent to traditional livelihoods. It mentions the methods used to move nomads and farmers and considers the adequacy of one-off compensation payments, unlikely to cover the long-term economic impact of mining on traditional livelihoods.

The fourth section examines Tibetan reactions to the above issues and the ways in which local communities have attempted to express their concerns over mineral exploitation, including petitions for compensation, and the direct action some Tibetans have taken to make their voices heard. It also provides examples of the responses of the authorities, which have ranged from the awarding of compensation to criminal prosecution, and notes that anger over Chinese mining activities has also added to general dissatisfaction with Chinese rule, and sometimes been an element in political protest in Tibet. Although many of the concerns that Tibetans have expressed over mining have been pragmatic, opposition to mining also arises from traditional religious and cultural beliefs and the relationship of Tibetans to their lands and waters. The last section on sacred sites considers this issue, which has been extremely sensitive for governments and mining companies throughout the world.

Constitution of the People's Republic of China (1984)

(Amendments Adopted at the Second Session of the Ninth National People's Congress and Promulgated for Implementation by the Proclamation of the NPC on March 15, 1999)

Article 9

- All mineral resources, waters, forests, mountains, grasslands, unreclaimed land, beaches and other natural resources are owned by the state, that is, by the whole people, with the exception of the forests, mountains, grasslands, unreclaimed land and beaches that are owned by collectives in accordance with the law. [...]

Article 10

- Land in the cities is owned by the state.

- Land in the rural and suburban areas is owned by collectives except for those portions which belong to the state in accordance with the law; houses' sites and privately farmed plots of cropland and hilly land are also owned by collectives.

- The state, may, in the public interest, requisition land for its use in accordance with the law. [...]

Ownership

According to the constitutional and legislative framework of the PRC, there are no 'Tibetan' resources. China's constitution places all mineral resources, waters and urban land under the ownership of the state, while grasslands, forests and mountains, as well as reclaimed, suburban and rural land are under state or collective ownership.[1] The State Council exercises ownership of mineral resources and other state-owned resources and land on behalf of the state.[2]

The state's ownership of mineral resources means that it also has the rights of occupation, utilisation, earning and disposition – rights, according to the Chinese Ministry of Land and Resources, that 'have nothing to do with the specific owners and users of the land covering mineral resources.[3] This means that local people are granted no control over and have little opportunity to benefit from mineral resources.[4]

1 Articles 9 and 10, see text box, opposite The system of collective ownership of land was inherited from the commune system established during the 1950s and 1960s throughout China with the abolition of private ownership and the collectivisation of land. The rural peoples' communes consisted of three levels: the commune, production brigade and production team. According to the revised draft of the Work Regulations for the Rural People's Communes (1962), the production team was identified as the owner of land within its limits (Peter Ho – see below). When the communes were disbanded in the early 1980s, they were replaced by farmers' collectives consisting of three levels that generally replaced commune, brigade and team: the township, the administrative village (*xingzhengcun*) and the natural village (*zirancun*) or villagers' group (*cunmin xiaozu*). However, there remains considerable ambiguity over the rights to collective land ownership among these three levels, with the middle level (administrative village), rather than the lowest level, usually acting as representative of collective ownership under the supervision of the township. See Peter Ho, 'Who Owns China's Land? Property Rights and Deliberate Institutional Ambiguity', The China Quarterly, Issue 166, SOAS: London, June 2001, pp.394-421. For an example of the subordination of the administrative village to the township in Yunnan province see Xiaolin Guo, 'Land Expropriation and Rural Conflicts in China', The China Quarterly, ibid, pp.422-439

2 Mineral Resources Law (MRL – amended 1996), Article 3 and the Land Administration Law (amended 1998), Article 2. There is no provision in the Constitution to determine who exercises state ownership of land and resources. It was not until the most recent revisions made to the Land Administration Law in 1998 and the Mineral Resources Law in 1996 that provisions were laid down in law to determine the representative of the state in exercising land and resource ownership. The revision to the land law acts to centralise land ownership and was introduced to 'overcome local protectionism' ('An Interpretation of the Land Administration Law of the PRC,' Beijing: *Falu chuban she*, 1998, p.68, in Peter Ho, op cit), resolving a situation where local governments, acting as representatives of the state had been exercising ownership. The same is likely to be true of the amendments that were made to the Mineral Resources Law. In the case of land administration, the revised law clearly indicates that it is the Ministry of Land and Resources (MLR) that is charged, under the State Council, with management and supervision of land, not local governments (Peter Ho, op cit).

3 MLR, 'China's Legal System for Management of Mineral Resources', undated, published (in English) on the MLR website www.mlr.gov.cn

4 It should be noted that the concept of centralised ownership of resources existed before Chinese Communist control over Tibet. The traditional Tibetan Government (Ganden Phodang) owned the whole country and therefore all natural resources. The government issued official documents permitting land use in return for taxes, payable in general to a governor installed by Lhasa, but sometimes directly to the central Tibetan government. In the case of noble families and monasteries, the situation was a bit different: they were also granted documents allowing them to use the land, but were exempt from taxes to a large extent. Moreover, noble families and monasteries collected taxes from the commoners, or 'serfs', who worked on that land and who were themselves considered to be a resource. In the parts of Kham not under Lhasa administration, the system, although varying from time to time and place to place, was the basically same: the ruler of a district owned everything and granted the right to land and resource utilisation.

The limited control that communities exercise over collectively owned land can be overruled by the interests of the state, as defined by the Party. The Constitution stipulates that the state can requisition collectively owned land for its own use if this is in the public interest (Article 10).[5] Given that the Chinese Communist Party exercises control over all state organs and that Party policies and interests precede and shape those of the state, it is effectively the Party that has the power to determine what is or is not in the 'public interest'. This can include the expropriation of land for mineral exploitation.

National autonomous areas are required by law to manage and protect natural resources within their territorial jurisdiction and are granted the right to exploit and use those natural resources that local authorities are entitled to develop.[6] They have no rights to ownership,[7] but are legally bound to support state exploitation.[8] A new clause (Article 65, Clause 2) has been introduced in the revised Regional National Autonomy Law (2001) stipulating that:

> *"The State should adopt measures that give a certain level of compensation to national autonomous areas from which natural resources are exported."*

The language of this clause is very vague and allows for wide interpretation. Without further clarification (and the establishment of mechanisms to monitor implementation) the actual benefit that autonomous areas will receive from the exportation of resources remains unknown. However, this does at least appear to be the first acknowledgement that autonomous areas should be compensated when resources are exploited by the state and transported to other parts of China, rather than being ploughed back into local development.[9] This is not a right recognised in the Chinese Constitution.[10]

5 This is dealt with by the 'Regulations Concerning Land Requisition for State Construction'.

6 PRC Law on Regional National Autonomy (1984; amended 2001).

Article 28 In accordance with legal stipulations, the organs of self-government of national autonomous areas shall manage and protect the natural resources of these areas.

[...] In accordance with legal stipulations and unified state plans, the organs of self-government of national autonomous areas may give priority to the rational exploitation and utilisation of the natural resources that the local authorities are entitled to develop.

7 In its General Recommendation No. 23 on Indigenous People's (1997), the UN Committee on the Elimination of Racial Discrimination called on state parties to the International Convention on the Elimination of all forms of Racial Discrimination (which includes China): 'to recognise and protect the rights of indigenous peoples to own, develop, control and use their communal lands, territories and resources'.

8 For example, PRC Law on Regional National Autonomy Articles 7 and 25. See Chapter 6, p.175-95 Recent amendments to the autonomy law (2001), bring the law into line with the drive to develop the western regions of China, introducing provisions to increase state control over and facilitate resource exploitation. See 'National autonomy law revised to support Western Development', TIN News Update, 13 March 2001

9 & 10 See footnotes on next page

In practical terms, the fact that the state has the sole rights to occupy, use, earn from and dispose of all mineral resources within the PRC's territory, does not necessarily mean that the state will exploit the resources itself. It also has the power to authorise occupation and utilisation of land[11] and resources by a third party (for example a domestic mining company), while the state benefits through fees and taxes. Indigenous inhabitants of locations where mineral deposits are found are subject to the will of either the state or such a third party, and often a combination of both.

Although the rights to own, use and benefit from resources should theoretically be exercised by relevant departments and according to state rules and regulations, the picture is more complicated at a local level where resource exploitation is often outside central control and has become subject to local protectionism. Although some Tibetans have benefited from decentralised, often technically illegal, exploitation, it generally leaves local inhabitants vulnerable to arbitrary expropriation and misuse of the pasturelands that they rely upon for their livelihoods. A foreign mining official whose company has sent geologists to prospect all over China mentioned the lack of control over gold mining:

> *"In some cases, as soon as a rich site is found, local officials, public security or the military simply take control of it. Therefore, an inherent conflict exists over who has the right to these resources on a local basis, regardless of what the State Council rules and regulations say."*[12]

In a few areas, Tibetan communities and their leaders do appear able to exert some influence over the development of mining at a grassroots level. More often, individuals have gained short-term benefits from mining, these might include cash compensation, employment, infrastructure improvements and wealth generation in their localities, although these benefits are largely unsustainable.[13] However, reports from Tibetans who have left China (either temporarily or permanently) suggest that most Tibetans feel detached from mineral and other industrial development, which is generally perceived as relevant only in so far as it impinges on traditional Tibetan ways of life.

9 (See previous page) International law recognises the rights of indigenous peoples to benefit from the use of resources. See for example International Labour Organisation (ILO) Convention (No. 69) concerning Indigenous and Tribal Peoples in Independent Countries (which China has not ratified). See Chapter 6, note 45

10 (See previous page) The Constitution simply states that: "*In exploiting natural resources and building enterprises in the national autonomous areas the state shall give due consideration to the interests of those areas*" (Article 118). The Mineral Resources Law (1986; amended 1996) states that: "*In mining mineral resources in nationalities' autonomous areas, the state shall give due consideration to the interests of those areas and make arrangements favourable to the areas' economic development and to the production and livelihood of the local minority nationalities.*" (MRL 1996, Article 10; MRL 1986, Article 33).

11 State owned land or land expropriated from collectives.

12 'Gold Mining in China: Taming the Wild West', US Embassy, China, May 1996. As discussed in the previous chapter, this kind of local protectionism has prompted the state to tighten central control over natural resource management. One of the amendments to the MRL adopted in 1996 was to define the State Council as the organ exercising state ownership of resources. See note 1

13 See Chapter 6

The lack of genuine participation in decision-making, or even cursory consultation, makes it virtually impossible for the average Tibetan to have any say over or involvement in development plans. The Chinese state has the power to allow anyone to move on to what Tibetans consider to be Tibetan land – and to force anyone to relocate. On a local level, officials (Tibetans and Chinese), the military or the police, can use and abuse their respective powers to override local concerns and to further their own interests. Even where local leaders are genuinely concerned with the well-being of their community, their hands are often tied by pressures from a higher level of authority. Such pressures can be the result of central policy as handed down through various levels and the need to meet set quotas; or can come from higher officials who seek to personally gain from a particular official or unofficial project.

Young Tibetan women work as gold miners in Dartsedo (Ch: Kangding) county, Kardze (Ch: Ganzi) TAP
© TOTAR, 1997

The average Tibetan nomad or farmer has little protection against the state and its centralised plans for large-scale resource exploitation, or against the corruption of local officials with vested interests in a particular mining project. China has been developing legislation to further codify mining and land use rights. However, this legislation has been drafted to protect the interests of the state vis-à-vis small-scale and artisanal mining or to protect the interests of potential investors – in particular foreign investors – who expect to be able to enjoy a minimum level of security in their investments.

State ownership of land and mineral resources is not peculiar to China. Many other countries also assert the state's right to land and resources under their respective constitutions, including Australia[14] and Indonesia[15] – both countries that have experienced conflict with indigenous populations over use of land. The extraction of mineral resources remains one of the most politically sensitive of land issues throughout the world.

The Chinese presence in Tibet bears a striking resemblance to Western colonialism, justified as it is by the ideological undertaking to 'liberate' and 'civilise' its 'backward' peoples, but motivated by clear political and military aims, with the principle economic interest in occupation lying in the potential riches of natural resources. Under the current regime, China's assertion of sovereignty over Tibetan areas has also meant an assertion of ownership over all Tibetan resources. Opposition to resource exploitation that poses a direct challenge to China's ownership of these resources can therefore also constitute a threat to the Chinese Communist Party's claim of legitimacy in controlling these areas, and even to the historical claims to the legitimacy of Chinese sovereignty over Tibet. The issue of resource ownership and mineral exploitation in Tibetan areas can therefore be both sensitive and deeply political.

14 Mineral resources belong to the state in most of Australia, although there are a few exceptions in states like Tasmania. According to the Australian Bureau of Statistics: 'Mineral resources are owned by the Crown in Australia, either by the state and Territory Governments within their borders (and up to three nautical miles offshore), or by the Commonwealth Government in offshore areas outside the three nautical mile limit. Accordingly, royalties are collected by state and Territory Governments for mining onshore and up to three nautical miles offshore, and by the Commonwealth outside that area.' ('Mining: administrative and financial arrangements', published on the website of the Australian Bureau of Statistics, www.abs.gov.au)

15 Constitution of the Republic of Indonesia, 1945 – Article 33, clause 3 The land, the waters and the natural riches contained therein shall be controlled by the state and exploited to the greatest benefit of the people.

Immigration

Many Tibetans see the immigration of large numbers of Chinese into Tibetan areas as the most serious threat to their land and resources and to traditional Tibetan livelihoods and culture. However, since the 1950s, the Chinese authorities have viewed the lack of human resources in Tibetan and other 'minority nationality' areas as one of the obstacles to developing the mineral resources of the Tibetan plateau. As a result, one of the associated impacts of the development of the mining industry has been the immigration of Chinese into Tibetan areas.

Before the foundation of the People's Republic of China (PRC), there was minimal Chinese presence in most Tibetan areas and the Chinese central authorities made no concerted attempt to exploit Tibet's mineral wealth. However, certain areas of Tibet were well known in China and elsewhere (for example, British India) for their plentiful gold resources – partly through direct experience and partly as a result of myth building. Chinese gold prospectors did sometimes penetrate Tibetan areas, particularly those bordering the Chinese world. For example, Chinese gold miners have been going to the area currently designated Chagzam (Ch: Luding) county, Kardze (Ch: Ganzi) Tibetan Autonomous Prefecture (TAP) since the late Qing dynasty. Mili (Ch: Muli) Tibetan Autonomous Ccounty (TAC) also now incorporated into Sichuan province, attracted gold prospectors who were sometimes killed by local Tibetans as intruders.[16] Chinese miners have been working gold and silver mines in Gyalthang (Ch: Zhongdian), Yunnan, since the 18th century.[17]

With the founding of the PRC and China's assertion of control over Tibetan areas, came the state's assertion of ownership over all mineral resources. The exploitation of Tibet's mineral resources was worked into national development plans. In August 1957 Zhou Enlai made a key speech on the incorporation of non-Chinese regions into the national plan, stating that it was not possible for the Han people alone to "*cast off poverty and backwardness and build a modern, powerful socialist state*". He pointed out the shortage of land and underground natural resources in the Han inhabited regions and the importance of developing natural resources in areas populated by the "*fraternal minority nationalities*" to support industrialisation. He went on:

> *"However, these natural resources have remained untapped for lack of labour power and technological expertise. Without mutual assistance, especially assistance from the Han people, the minority people will find it difficult to make significant progress on their own".*

16 'Survey of Mili TAC', p.15, in Steven D. Marshall and Susette Cooke, 'Tibet Outside the TAR', 1997
17 'History of Tibetan Society in Diqing', p.187, Marshall and Cooke, 1997, op cit.

During the first thirty years of the PRC, this shortfall in labour power was met to some extent by the forced colonisation of some of the more sparsely populated areas rich in resources, such as Qinghai province, Inner Mongolia Autonomous Region and Xinjiang Uyghur Autonomous Region (XUAR). The population of the labour camp network[18] and the military also played a role in the construction of infrastructure and settlements and the development of resources.

During the last two decades, improvements in infrastructure and policies to encourage the voluntary migration of Chinese to Tibetan areas have provided a flow of labour to the developing mining industry. This has been most notable in the Tsaidam Basin, Qinghai, but in many other Tibetan areas there has been considerable growth in the number of smaller-scale mining enterprises, both official and unofficial, attracting large numbers of migrant workers and prospectors. Added to this have been the tens of thousands of artisanal miners, concentrated mainly in Yushu TAP in Qinghai, Nagchu (Ch: Naqu) prefecture in the TAR and Kardze TAP in Sichuan.

Mining town near Nakartse (Ch: Langkazi), Lhoka (Ch: Shannan) prefecture **© TIN**

18 See Chapter 4, p.139-142

Large-scale mining and industrialisation

Medium to large-scale mining enterprises depend upon access to large numbers of unskilled and skilled labourers for infrastructure construction and mineral extraction. They often require the development of new settlements for the medium to long term, depending on the life of a particular mine and on the existence of other exploitable reserves in the area. If the majority of workers are Chinese who have been brought into the area to provide labour, this can significantly alter the demographic balance in a particularly locality. When several such enterprises are grouped together the impact is compounded, as has been the case already in Tsonub (Ch: Haixi) prefecture (see below).

Depending on the life of a mining enterprise and other opportunities in the area, workers may bring in families and settlement can be over more than one generation. In some cases their residency (*hukou*) may travel with them, they will become part of an area's official population and settlement will be permanent. An infrastructure and services are necessarily built up to provide for these families, forming what is effectively a small town. For example, a former employee of the Tsakha (Ch: Chaka) salt factory in Wulan county, Tsonub TAP, described the settlement that has been built up for workers:[19]

> *"There are about 1,000 workers in the salt factory. There aren't many local people. Most are Chinese. They have a place to stay in the factory. In this factory there is a public security office, a hospital and a school. Then there is a crèche and a cinema hall. There are about 300 students [in the school], up to higher middle-school standards. If they get through the examination they can go to university, If they don't pass the exam they will be given work in the factory through the personnel office. The school was built solely for the factory, therefore the people of the area can't go to that school."*[20]

In some areas, Tibetans claim that infrastructure (power, water, roads), housing and the provision of services (health, education, leisure) in state-owned mining enterprises are superior to those in place for indigenous inhabitants. This inevitably leads to resentment from the local population.

19 Chinese settlement of the Chaka district goes back to the early years of the PRC. Forced immigrant settlements and agricultural *laogai* had been established at Wulan's county seat and in the Chaka district during the 1950s. Marshall and Cooke, 1997, op cit.
20 TINDoc 7(rp)

Immigration and mining in the Tsaidam Basin

During the last fifty years, the Tsaidam Basin in Tsonub TAP, Qinghai, has developed into the most important area for mineral resource exploitation on the Tibetan plateau. Prior to the 1950s, the area was sparsely populated by a predominantly Tibetan and Mongolian nomadic population. The early development of both agriculture and industry in Tsonub was largely dependent on the forced resettlement of population from inland China, the 're-education through labour' of prisoners and intellectuals who were put to work in the sprawling labour camp (*laogai*) network and the military.[21]

As early as 1953 the Chinese government had dispatched geological survey teams to explore the Tsaidam Basin. As a result of their findings, the central authorities decided to give priority to the development of industry in Tsaidam over the development of pastoral areas.[22] Oil prospecting began in 1955 and the construction of settlements in the mining areas of Dachaidan and Mangya began in 1956.[23] The first immigrants to arrive in Tsonub in the early 1950s were Chinese prisoners, soldiers and guards. These were followed by thousands of forced immigrants deported for resettlement from the eastern regions of China. Forced immigration intensified at the end of the 1950s, and included central Tibetans who were sent to Qinghai's labour camps following the 1959 Tibetan uprising. From 1956 to the mid-1960s city dwellers were sometimes transferred from eastern and central China and until the 1970s excess population was also transferred from Xining.[24] Many forced immigrants and ex-prisoners remained in Qinghai as permanent settlers. Over the past 20 years, improvements in infrastructure, most notably the opening of the Xining-Golmud railway in 1984, and policies to encourage voluntary immigration have accelerated the development of mining areas and the colonisation of Tsonub.

According to an article in the journal 'Tsaidam Development Resarch', the Tsaidam Basin now has 32 urban centres, including the cities of Golmud and Terlenkha (Ch: Delingha). Most of the construction and development of towns in the basin has been related to mineral products (for example the oil town Lenghu).[25] Many of these mining towns, including Mangya and Dachaidan, were originally founded on the labour of forced immigrants and prisoners.

21 In the 1970s, the PLA dismantled some of the *laogai* organisation and took over key projects such as the Golmud-Lhasa petroleum pipeline and the Xining-Golmud railway. James D. Seymour and Richard Anderson, 'New Ghosts, Old Ghosts' , New York and London: M.E.Sharpe, 1998

22 June Teufel Dreyer, 'Ch'inghai', in Edwin A.Winckler (ed.), A Provincial Handbook of China, Stanford University Press, c.1975, cited in Marshall and Cooke, 1997, op cit.

23 Marshall and Cooke, 1997, op cit.

24 Marshall and Cook, 1997, op cit.

25 Cun Wanjuan and Zhang Zhimin (Qinghai Provincial Hydropower Department), 'Water resources and the construction of small towns in Tsaidam', Tsaidam Development Research (*Chaidamu Kaifa Yanjiu*), Vol.1, 2001

The western part of the basin, with Golmud as the industrial centre, has developed around the exploitation of the area's salt lake resources, petroleum, non-ferrous metals and construction materials. The eastern part of the basin, with the prefectural capital Terlenkha as its base, provides an agricultural base – founded on the *laogai* system[26] – for the prefecture's industry and also has its own salt industry.

Although Tsonub prefecture retains it Mongolian and Tibetan autonomous status, 2000 census figures indicate that the population is now 65.6 per cent Han, 13.1 per cent Hui, 11.1 per cent Tibetan and only 6.7 per cent Mongolian.[27] The number of Chinese not included in this official figure is likely to be significant, particularly in Tsonub's main urban centres, Terlenkha and Golmud. Both are located on the railway line and are centres of mining and industrialisation and are therefore likely to attract a large floating population.[28] Over one third of the Tibetan population of Tsonub is concentrated in Themchen (Ch: Tianjun) county, in the east of the prefecture, which has the smallest population of all Tsonub's counties. The major mining and industrial urban centres are predominantly Chinese. For example, according to the 1990 census, the combined populations of the key mining towns of Dachaidan, Mangya and Lenghu (a total of 73,300 residents) were 95 per cent Chinese.[29]

26 The first prison camp – an agricultural *laogai* – to be established in Qinghai was in the then 'barely populated place called Delingha'. In 1966 the administrative seat of the prefectural government was moved there due the availability of food. It was declared a town (Ch: *zhen*) in 1973 and promoted to city (*shi*) status in 1983. In mid-1986 Delingha was functioning as a normal city administration. In 1987 the agricultural *laogai* became a state farm, the Qinghai provincial *Laogai* Bureau had lost its best farm, and prisoners who hadn't died or been released were transferred to other camps. Many former prisoners remain in the area. James D. Seymour and Richard Anderson, 1998, op cit.

27 Xue Zheng (ed), Qinghai Province Bureau of Statistics, 'Qinghai Statistical Yearbook, 2001', China Statistics Press, Beijing 2002. Although 2000 census data shows little growth in the official Han Chinese population of Tsonub during the last ten years (only 5,000 more than 1990), the Hui population has trebled, from 15,615 according to 1990 census data (representing 5 per cent of total population), to 48,497 (representing 13.1 per cent of total population). As a result, according to official data there are now more Hui in Tsonub than Tibetans, most of them concentrated in Golmud.

28 According to the Tsonub Annals (*Haixi zhou zhi*), published in 1995, the yearly floating population for Terlenkha municipality had reached 25,000 people – half as many people as those who were registered as official population (about 50,000). Marshall and Cooke, 1997, op cit.

29 For more information on the demographic restructuring of Qinghai province, in particular Tsashar (Ch: Haidong) and Tsonub prefectures, see Susette Cooke, 'The politics of population transfer', TIN Special Report, 28 October 1999. See also Marshall and Cooke, 1997, op cit.

Small-scale mining

During the last two decades, immigration into Tibetan areas has largely been a result of individual economic migration.[30] Since the 1980s, economic liberalisation and policies introduced to ease restrictions on population movement have resulted in more settlers and migrant workers arriving from other parts of China. The far-reaching economic reforms instituted following Deng Xiaoping's tour of southern China in 1992 resulted in a large increase in the number of migrant workers looking for employment. Several factors have maintained this flow of migrant labour throughout China. These include an easing of restrictions on population migration, the huge surplus of under-employed rural labour in China[31] and the growing unemployment rates in urban areas as State Owned Enterprises (SOEs) continue to lay off workers. While the majority of these migrants have headed to the eastern coastal areas, there has also been an increase in the numbers of Chinese travelling west to Tibetan areas in search of work, particularly those from densely populated neighbouring areas, such as Sichuan province and Xining and Tsoshar (Ch: Haidong) prefecture in Qinghai.

As government departments and organisations, collectives, SOEs, the military and the police were encouraged to join in with the new entrepreneurial spirit of liberalisation, small-scale mining operations were among those projects embarked upon, and local governments started leasing out land to private prospectors. In Tibetan areas, these small-scale mines generally rely on migrant labourers to make up at least some, but often the bulk, of their workforce. Mines that operate on a relatively short time-scale do not usually involve the construction of housing for workers – labourers usually stay in tents. Once a lease has run out (which can be as quickly as a year for small-scale private miners), most of the in-comers will go elsewhere in search of new work, although some may settle.

Particularly apparent are the large numbers of artisanal miners panning or digging for gold in Kardze TAP in western Sichuan, Nagchu prefecture in the TAR and Yushu prefecture in Qinghai. For example, Steven Marshall and Susette Cooke, who carried out extensive fieldwork in Tibetan areas outside the TAR in the mid-1990s, described the heavy concentration of prospectors, mainly Chinese, in Dege county, Kardze TAP. These miners were working along the river gorge leading to the county town, sleeping and working in makeshift camps along the riverbed:

30 Although most migration throughout China is now voluntary, the authorities are implementing official resettlement projects as part of official poverty alleviation schemes. This includes the resettlement of approximately 58,000 poor farmers, mostly Chinese and Hui Muslim, from Tsoshar to Dulan county, Tsonub M&TAP. Although the project is no longer being funded by the World Bank (it was originally Component C of the Bank's China Western Poverty Reduction Project), China is proceeding with the project on its own terms. See 'Resettlement and urban construction in former World Bank project county', TIN News Update, 14 February 2002.

31 According to AsiaWeek, the Chinese government has said that half of its 900 million peasants are underemployed. Allen T. Cheng, 'A Rural Dilemma: While WTO accession will not lead to chaos, China still has its work cut out', AsiaWeek, 23 February 2001, Vol.27, No.7

"Between the base of the Cho-la pass and the county town, at least 1000 miners would be working, even early in the season. The smallest sites were worked by six to ten miners and were often Tibetan. The largest had over 20 miners, up to 30 or even 50, and were generally Chinese. They live in cold, squalid conditions, under sheets of reinforced plastic crudely made into tents, along the river's edge amidst the gravel. Most are young, and some wives accompany their husbands to these remote, uncomfortable places." [32]

Chinese gold-miners' camp between the Cho-la pass and Dege county town, Kardze TAP © TOTAR, 1997

32 Marshall and Cooke, 1997, op cit.

One of the Tibetan areas affected most by mass artisanal mining has been Chumarleb (Ch: Qumalai) county in Yushu TAP, Qinghai.[33] According to an article in the Qinghai Daily in January 1999, thousands of 'gold peasants' flooded into the county from 1982 to dig for gold. The article states that at the busiest times there were more than 30,000 prospectors at the main gold fields. Despite official efforts to curb illegal mining in Yushu prefecture in the 1990s, the presence of large numbers of immigrant workers was still obvious at the end of the decade. The article reports that following an official clampdown on illegal mining, the local authorities in Chumarleb county leased land out to private workers for small-scale enterprises:

> *"Panning beds lie side by side along the 70 to 80km of riverbank by the road through Dongfeng and Ganba villages in the west of Chumarleb county. Bulldozers and high-pressure water machines puff out gritty air as they work flat out, while nearby the grasslands are covered with the white tents where the workers live. Judging from the number of tents, there are at least 100 people at each mining spot and there are nine such places altogether."*[34]

The presence of at least 900 quasi-official mine workers along one 70-80 km stretch of riverbank, let alone the reported tens of thousands of illegal miners who had mined the area beforehand is particularly striking given the area's resident population. According to the 1990 census the official population of Chumarleb county, which has an area of 48,000 square km, was only 18,983 people.[35]

The thousands of artisanal miners who have entered Tibetan areas to prospect for gold and live in makeshift camps, are the least likely to settle permanently, although they may move on to other Tibetan areas. The regions that have been hit worst by 'gold fever' in Qinghai and northern TAR are remote and at high altitude and, apart from relatively small urban centres, they are still predominantly inhabited by nomads. Although artisanal miners may have less of a long-term official demographic impact on the area which they mine, they have already caused lasting environmental damage in certain areas. This affects the livelihoods of herders dependent on the grasslands that have been mined.[36]

33 See also Chapter 5, pp. 171-179

34 'Exploring the corridor of Gold', Qinghai Daily, 18 January 1999

35 According to 2000 census data the population has risen to 24,769, with 97.5 per cent Tibetans and only two per cent Chinese, generally reflecting the official ethnic breakdown for Yushu as a whole – over 97 per cent Tibetan, 2.2 per cent Chinese (Qinghai Statistical Yearbook, 2001). Although the Chinese population is likely to be higher than official statistics allow, according to Marshall and Cooke, Yushu is still the 'most Tibetan of Tibetan all regions' due to remoteness, high elevation, cold temperatures, lack of land suitable for farming (most of the population are nomadic pastoralists) and poor infrastructure. By the mid-1990s, Chinese and Hui Muslims did not represent a particularly large or noticeable presence even in the prefectural capital of Kyegudo (Ch: Yushu). This was expected to change with the improvement of infrastructure linking Yushu to Xining. (Marshall and Cooke, op cit. pp.2370-2378)

36 See Chapter 5

Tibetans generally resent the presence of large numbers of Chinese workers.[37] Similar to residents in the coastal cities of eastern China who resented the influx of rural migrants during the 1990s, they tend to blame immigrants for increasing socio-economic problems such as rising crime and unemployment. However, in Tibetan areas Chinese immigrants have the advantage of language and often skills, and Tibetans are finding it increasingly difficult to compete in the job market. There is also a cultural dimension to antagonism towards Chinese immigrants. There is a fear that greater numbers of Chinese will inevitably lead to loss of Tibetan identity and culture as their local environment becomes ever more Sinicised. Added to this is a sense of powerlessness and anger over the exploitation by Chinese workers of what are considered by Tibetans to be Tibetan resources, exacerbated by the fact that it is the Tibetan farmers and nomads who are left to cope with the damage done to their lands and waters after miners have left.

In 1994 a large gold vein was discovered in Nagchu prefecture, TAR. When this was reported in the press, large numbers of migrant workers, many of whom were Hui Muslims from neighbouring provinces, came to the area to exploit the reported riches.[38]A Tibetan from the area interviewed at the end of 1994 said that the influx, which began in mid-June, increased dramatically in October and that the area was "*gradually being taken over by Chinese migrants*". This feeling of the area being taken over by prospectors and migrant workers is also apparent in the following account of two brothers, both in their sixties, from the same area of Nagchu. The negative comments that they make about the Hui immigrants are in part based on their personal experiences of competition for work and land. However, they also reflect a more general attitude held by Tibetans towards the Hui, who are well-known for their entrepreneurial skills and ability to survive and make a living in difficult circumstances.[39]

> *"There are lots of people coming by trucks to live in new settlements around where we live. They take Tibetan resources, very expensive precious stones and gold. Before this place was empty, except for animals. But now it is full of these Chinese who are Muslims [Hui]. They are famous businessmen. They have robbed lots of sheep from Tibetan nomads. I cannot even measure the number of Chinese coming into Nagchu. They're there until the resources run out."*[40]

37 It should be noted here that Tibetans often include Hui Muslims when referring to 'Chinese'. The Hui are often referred to as Chinese Muslims because they use Chinese language. However, they consider themselves to be ethnically different from the Han Chinese. The Hui are largely concentrated in the north-west provinces of Shaanxi, Gansu and the Ningxia Hui Autonomous Region.
38 Asahi Shimbun, 8 May 1995
39 There were pockets of Hui settlers in the north-east of Amdo pre-1949 and a small Hui community in Lhasa. However, over the past 20 years, the numbers of Hui that have come into Tibetan areas (most notably the areas of Qinghai beyond Tsoshar/ Xining) as economic migrants, setting up shops, restaurants and businesses, has fuelled Tibetan antagonism towards them.
40 TINDoc 35(sd)

Population control

> *"Not a single country or region can bear unlimited population growth. Even if a region enjoys a large area and rich natural resources, it has limitations. A limited space cannot contain unlimited population."*
> Huang Rongqing, director of the Institute of Population Economics, Capital University of Economics and Business, Beijing.[41]

The influx of large numbers of mine workers, their impact on the local environment, their drain on local resources, and in many cases their different cultural and social traditions, have been sensitive issues for mining companies working in many countries. This is particularly the case where resources are being developed in areas with an autochthonous population of distinct ethnicity to the dominant nationality of the governing state. The resulting marginalisation of the local community, both economically and socially, has often led to disputes and ethnic tension.

Chinese gold miner awakes; a section of pump hose for mining lies in the foreground. North of Dege county town, Kardze TAP © TOTAR, 1997

41 Huang Rongqing, 'Population and Development of Minority Nationalities in the Western Region', Beijing Population and Economics (*Renkou Yu Jingji*) , 25 November 2001. Translated by FBIS, 11 January 2002

Reports in China's official press suggest that there is some understanding of the pressure that a growing population will put on the western regions of China, including Tibet,[42] principally focusing on the environment and drain on local resources. However, no connection is made in official reports between the immigration of Chinese to serve development aims, including infrastructure construction and resource exploitation, and the current strain on Tibet's and other western regions' resources. These official reports suggest that the sole cause of current population pressures is the relatively high birth-rate amongst the 'minority nationality' populations, and respond with calls to tighten control over family planning and improve the 'quality' of national minority populations.[43]

Factors influencing the demographic restructuring of Tibetan areas over the past fifty years that are ignored in this analysis include:

- the increase in population as a result of forced immigration before the 1980s, particularly in Qinghai province, Xinjiang Uighur Autonomous Region (XUAR) and Inner Mongolia;
- the voluntary immigration of Chinese from the 1980s as a result of economic reforms and incentives;
- and the growing 'floating population' of Chinese, including the tens of thousands of migrant workers who have entered Tibetan and other western areas since the 1980s to prospect for gold and hunt.[44]

The paradox in state policy is clear: the growth of minority nationality populations is to be limited to allow for sustainable development in western China, but at the same time, Chinese are encouraged to come to Tibetan areas to participate in resource exploitation among other economic development.

42 See for example Huang Rongqing (op cit), or China Daily 14 December 2002. The latter blamed environmental deterioration in the western regions on "*rapid population growth*", adding that "*In order to support the increasing population, forests and pastures have been turned into farmland, which has upset the region's ecological balance*".

43 The meaning of the term 'population quality' is often vague. It has been used to refer generally to educational, scientific and cultural levels, but has also been used specifically to refer to mental and physical health. See the essay 'Population, Migration and Birth Control' in News Review: Reports from Tibet 2000, TIN, February 2001. For a wider discussion of eugenics in China and Tibet see 'Survey of Birth Control Policies in Tibet', TIN Background Briefing Paper, 30 March 1994, pp. 27-30

44 Among Tibetan areas, the added population pressure from Chinese settlers is particularly notable in Qinghai province. Most of Qinghai province was traditionally inhabited by nomadic herders, with farming largely limited to the eastern edges. The only part of Qinghai with a majority Chinese population prior to 1949 was in the north-east, in the area which now constitutes Xining municipality and Tsoshar prefecture. Although some small settlements of forced immigrants were established outside the Tsoshar/Xining zone during the Republican period, the population was still overwhelmingly Tibetan and Mongolian (Cooke, 1999, op cit). By contrast, according to 2000 census data, almost one third of the combined official populations of the six Qinghai prefectures outside Xining/Tsoshar, all of which have Tibetan autonomous status (Tsonub M&TAP, Tsojang (Ch: Haibei), Malho (Ch: Huangnan), Tsolho (Ch: Hainan), Golog (Ch: Guoluo) and Yushu TAPs) is Han Chinese (493,318 out of a total 1.68 million; representing 29 per cent) and Hui Muslim (98,312 representing 6 per cent). The Chinese population is likely to be considerably higher than allowed by these official statistics, which do not include the military or floating population of unregistered migrant workers.

Land expropriation

One of the greatest impacts on nomads and farmers in areas marked for mineral exploration and exploitation is the loss of pasture and arable land. Water resources, also under state ownership, are often used to provide power and water for mines, and both ground and surface water is at risk of contamination from mining and mineral processing enterprises.[45] Without adequate compensation mechanisms and the provision of suitable alternative resources, the expropriation of land and water for mineral development, and subsequent environmental damage, can pose a threat to the sustainability of the traditional livelihoods of rural Tibetans.

Tibetan nomads in some areas have lost significant tracts of pastureland to the various forms of mining enterprise, from the larger state-owned enterprises to small-scale and artisanal miners. In the rural areas where exploitation has taken place, mining has added to other pressures on local resources resulting from a growing population and the degradation of the fragile pastureland of the Tibetan plateau. Where there has been extensive gold mining, nomads are not only temporarily denied access to grazing land, but the land is also frequently unusable once the miners have left. Larger mining enterprises that involve the employment of hundreds of workers require housing and other associated urban construction that further encroaches on arable and pasture land.

The awarding of compensation for loss of land appears to vary from area to area and from case to case. Compensation payments are sometimes automatically assigned and paid; other reports indicate that locals sometimes have to petition or argue for compensation before they receive it.[46] In some cases Tibetans have reported that they have been promised compensation but have never received any. In other cases, compensation has been paid but is deemed inadequate to cover loss of livelihood. There are no mechanisms in place to compensate those who have suffered as a result of artisanal mining and the damage this has done to both grasslands and water sources.

45 See Chapter 5

46 For examples of action taken by locals in opposition to mining or to demand compensation see below p.101-110

According to a former government official, land has been expropriated from 40 nomad families on the site of Saishitang copper deposit in Tsigorthang (Ch: Xinghai) county, Tsolho TAP one of Qinghai's three major copper mines.[47] To compensate for the decrease in the area of their pastures, the nomads were given a cash payment by the state. However, one of the problems with one-off cash compensation payments is that they do not compensate for permanent loss of livelihood if the nomads or farmers are not given other land or are unable to find alternative employment. The shortage of land in some Tibetan areas makes this particularly problematic. A Tibetan from Rebgong (Ch: Tongren) county, Malho TAP, Qinghai, commented that there was simply no other land in his local area to give to nomads in compensation for use of local pastureland for government-run gold mining.[48]

A Tibetan from Jomda (Ch: Jiangda) county, Chamdo, talked about work being carried out in 2001 in Chunyido[49] (Ch: Qingnitong) village, located about nine km north of the Sichuan-Tibet highway (#317) where the Yulong copper mine is located.[50] This included the construction of a road through the valley and about 18 wooden houses. He says that according to the boundaries marked by the authorities 50-60 families will lose their summer pastures to the mine:

> *"According to the plans these grasslands will be under construction as part of the mining plan. If this grassland is in use by the mine, the nomads won't have any [summer] pasture. Although they measured the area and drew lines last year [2001], the site had not yet been fenced off, so the nomads could still go there with their animals. The nomads use it for three months each year during the summer. 50-60 families use this land and each family has about 60-70 animals."*[51]

47 See Chapter 1, p.31-32
48 TINDoc 554(jw)
49 Wylie: '*Chu nyid mdo*'
50 The TAR's most important mining site. See Chapter 1, p.31-32 including a map of the area. According to the TAR National Land Specialist Plan, the Qingnitong deposit, referred to as the 'Yulong mining area', is the first of the mining areas located in the Yulong mining zone that is to be developed. The Plan states: "*The mining area is purely an animal husbandry area, on average 4,500m to 4,600m above sea level, where herdsmen only turn livestock out in summer and autumn to graze in the mining area or nearby.*" The Yulong mining zone has a total area of approximately 1870 square km (400km long and varying between 30 and 70km wide). Four other deposits/mining areas in the Yulong mining zone were identified by the plan for supplementary prospecting work 2000-2010 (place names given in Chinese): Malasongduo, in Kuoda village, Chaya (Tib: Dragyab) county; Duoxiasongduo, in Seduo village, Xiangpi township, Gongjue (Tib: Gonjo) county; Zhanaka and Manzong mining areas in Tuoba township, Changdu (Tib: Chamdo) county. ('TAR National Land Specialist Plan [*Xizang Zizhiqu Guotu Zhuanti Guihua*], 1996-2020, TAR National Land Plan Formulation Committee [*Guotu Guihua Bianzhi Weiyuan Hui*] and the TAR Planned Economy Committee [*Jihua Jingji Weiyuan Hui*], internal [*neibu*] document, undated, pp.495-500)
51 TINDoc 749(jw)

This Tibetan was not aware of any plans to resettle the families elsewhere, although there will clearly need to be provision of summer pastures for their livestock. However, he thinks that this could be difficult due to existing pressure on available land:

> *"The only thing they can give in return for taking this land is the land in use by other Tibetans, so whatever [land] they given in return, it will inevitably cause grave problems."*[52]

Methods used by local authorities to move nomads and farmers whose land is needed for a mining enterprise range from persuasion, offering monetary or other compensation, to threats and sometimes a combination of both. Where the enterprise in question has the backing of higher levels of authority, local officials are obliged to gain compliance one way or another.

A Tibetan who used to work in a county-level administrative department in Qinghai, described what happened when he went to inform a group of nomads that they had to leave a site that had been marked for mining:

> *"The nomad families who used the land were made to leave the mining site. The nomads were told that they should listen to what they were told. Since I was an employee of a relevant office I had to go there to give advice to the public. I went there many times.*
>
> *"At first the nomads said that they would not hand over the land. However, they couldn't challenge the authorities and since they had no other alternative they said that they would agree to leave the place. They are not educated people and therefore it was difficult to talk to them, but they could say nothing and had to leave the place when the government directed them to do so.*
>
> *"I told them, 'This is a large construction [project] for the nation, and you should do what the government tells you to, if you don't, then you can be arrested. If you are arrested we won't be able to help you.' I had asked the county head and the Party secretary what I should do if the nomads didn't leave the place after we told them to vacate it. They told me that if they refused to leave the place I should arrest them. Public Security Personnel also went with me [to meet the nomads]."*[53]

52 Ibid.
53 TINDoc 695(jw)

He said that the nomads were given monetary compensation, distributed through the county's Land Administration Office, and described how the compensation system worked:

> *"The money that was paid to the nomads was paid through the Land Administration Office. The government gave the money to the office and they had to take the money and distribute it to the nomad families. First the office measured the area of the land to be exploited for mining and inspected how many people live on it. They reported the details to the higher levels who then calculated [the amount of compensation] and sent the money to the Land Administration Office. The office then gave the money to the public."*[54]

As well as the long-term expropriation of land for state and local government mining projects,[55] nomads and farmers have been adversely affected by local government initiatives to sell short-term leases to private gold miners. By the time the land expropriated for mining is returned to the original users it has often become unserviceable.

According to Chinese law, the responsibility for cleaning up a site lies with the unit engaging in mining operations. Available information indicates that whether a mine is run by the state, the local government, or private individuals, environmental standards are generally poor, and mining enterprises and local government officials have generally not held themselves accountable for difficulties that mining has caused local communities. As a result there has been widespread failure to restore sites after mines are closed.[56] Whilst this is a problem that is officially acknowledged, supervision from higher levels to ensure suitable restoration operations take place remains grossly inadequate. The environmental consequences of poor site management including land degradation and pollution, inevitably has a major socio-economic impact on families reliant on the land.

54 Ibid.

55 There have also been allegations that government forced resettlement schemes in Phenpo Lhundrub (Ch: Linzhou) county, Lhasa municipality and Gongjo (Ch: Gongjue) county, Chamdo prefecture, TAR, and in Nyagchu (Ch: Yajiang) county, Kardze TAP, Sichuan, are linked to plans for mineral exploitation in these areas. Although able to confirm that such resettlement initiatives are taking place in Phenpo and Gonjo, TIN has been unable to confirm the alleged links to mining. The official press has reported that 144 households, comprising 948 farmers and herders, are to be moved out of Gonjo and into Pome (Ch: Bomi), Nyingtri (Ch: Linzhi) and Miling (Ch: Milin) counties in Kongpo (Ch: Gongbu) prefecture, officially as part of a forestry protection and poverty alleviation project (tibetinfor.com 31 May 2002). The official explanation given to villagers from two villages in Ngangnang township, Phenpo Lhundrub, who were relocated within the area, was a health concern over the local water supply, although villagers say that they rely upon the same water supply in their new homes [TINDocs 12(se); 884(jw); 885(jw)]

56 See Chapter 5

Government, collective and private gold mining has been prolific in Lithang (Ch: Litang) county, Kardze TAP, over the past two decades. A gold production drive was started in the 1980s, when gold mining was put in the hands of village and township collectives.[57] By the 1990s, there was also at least one large government gold mining site located about 40 to 50km to the west of the county town.[58] Between 1995 and 1996, the Lithang People's Liberation Army base was converted into a mining support facility for the 'Kardze TAP Gold-factory-gully Gold Mine (*jin chang gou huang jin kuang*)'[59] and a new slogan had appeared inside the compound: "*Gold is truly beneficial, gold is truly life*".[60] In 1995 the county authorities also started expropriating farm and pasture land to lease to private prospectors who set up small-scale gold mining enterprises.

Although the prefecture is reported to have banned private mining in the county in 1999 because the situation had got out of control, poor (or non-existent) environmental management in both state and private mining areas had already led to extensive land degradation in the areas that had been mined. A young Tibetan from the area talked about the long-term damage done to the grasslands in Lithang as a consequence of mechanised extraction:

> *"By using machines for excavations for the last ten years or so, the whole pastureland has been turned upside down. Before gold mining started, the area used to be a vast pastureland popular with the nomads. Now the Chinese have turned the whole area upside down with their excavations and diggings. This is causing great hardship to the nomads and their animals."*[61]

The private gold mining activities in Lithang led to the expected tensions between prospectors and local farmers and nomads, reportedly leading to fighting in 1997. However, according to a Tibetan who rented land from the government for gold mining, disputes also arose among the different nomadic groups in the area over compensation payments for land expropriation. The land in this area remained assigned collectively to each nomadic group and had not yet been redistributed amongst family households. When the authorities offered compensation payments for requisition of the land for private mining, disputes broke out over who should receive such payments:

57 Kardze Annals Editorial Committee , 'Kardze Annals', Chengdu: Sichuan People's Publishing House, 1997, p.1180
58 TINDoc 446(jw)
59 The Kardze TAP Gold-factory-gully Gold Mine was constructed as part of Kardze TAP construction drive and was one of the prefecture's leading gold producers by 1989-1990. (Contemporary Ganzi, in Marshall and Cooke, 1997, op cit).
60 Marshall and Cooke, 1997, op cit.
61 TINDoc 23(tv)

> *"The problems arose because the government pays compensation for places where gold mining is done. Thus the different nomad groups fight over whose land it is in order to get compensation. If the Chinese have clearly identified and allotted land of all the groups to be mined, then there is no problem. If they only give out the land of one of these groups, or two of three groups [for mining], then there are many problems. Then all the groups will claim that the land concerned is theirs."*[62]

He states that this was particularly problematic where the land being mined was on, or close to, the border between nomadic groups. Whichever group was not being compensated would oppose the mining.

Digging the pastureland for gold, Kardze TAP **© Anders Højmark Andersen, 1994**

62 TINDoc 446(jw)

Protest

Under the current system in China, people are able to make complaints about land issues either directly by petition or through their local representatives to a higher level. Many of the Tibetans who have spoken to TIN about mineral exploration and exploitation in their areas have said that local people made representations to their local authorities (and sometimes directly to the mine authorities) to express their concerns or make complaints. These complaints have usually been a result of loss of pasture or arable land, damage to the local environment affecting livelihoods and health, and immigration. In many cases Tibetans are petitioning for some form of compensation – indicating that there is still a perception that they have a right to compensation from the government for loss of what they consider to be Tibetan land and resources.[63] Although some Tibetans have called for a complete cessation of mining activities, many appear to feel that requesting a halt to mining activities is futile, whereas requesting compensation is more realistic. However, anger over Chinese mining activities has also added to general dissatisfaction with Chinese rule over Tibet. As a result, protest against resource exploitation has sometimes been an element of the pro-independence demonstrations that have taken place since the late 1980s.[64]

Petitions and representations to the higher authorities, usually made by a local village or township leader, have been met with varying responses, as has the direct action which has been taken by some communities to disrupt mining activities (see below). The most positive responses have been payment of, or at least an undertaking to pay, some sort of compensation, while the most extreme response has been fixed-term imprisonment.

Some officials and community leaders at a local level (usually township and lower) are sympathetic to the concerns of their community and have sometimes attempted to prevent or halt mining activities, or at least appeal for compensation for the community. In one case, rather than going through official channels, a township head resorted to traditional methods of influencing events and invited lamas to the local area to give prayers for the Chinese not to extract minerals.

63 In the case of land expropriation this may be a valid assumption: if the land is collectively owned, the collective should be compensated by the state when its land is requisitioned. However, as mentioned above mineral resources are owned by the state; local people have no rights to be compensated for their utilisation.
64 The main issues prompting such protests, which are most commonly manifested in the form of demonstartions and/or postering/leafleting, have been political, religious and cultural repression and immigration. However, use and misuse of Tibet's resources has also prompted much anger, and calls for an end to deforestation and/or mining have been incorporated into more general opposition to Chinese rule.

However, when a project has been officially approved by higher levels and is viewed as important, local leaders can come under heavy pressure from higher levels to quash any opposition. It is the obligation of each level of government and all PRC citizens to place the interests of the state above all else, and to fulfil tasks assigned by state organs. As a result, any sign of unwillingness to welcome mining and other development can be construed as recalcitrance. During the mid-1990s, locals of one township in Nagchu prefecture had complained about the effects of mining on the local environment. However, the township Party secretary was effectively silenced by the county authorities when he attempted to voice the concerns of the local community:

> *"People voiced their opinion to the township Party secretary and he conveyed the message to the county administration. But the county administration said that the people were raising their voices because the township administration was not able to control them. They said that there must be leaders among the people instigating them to say all this, so if the protest continued they would arrest the instigators. People said that they were going to go to the TAR government to complain, but no-one had the courage to volunteer. People were afraid they would have problems later on."*[65]

In addition to subordination to the interests of the state, officials and citizens are also subject to the personal interests of officials at higher levels. The regular rotation of county and township-level officials can have a significant impact on a local community, particularly where a newly posted official is primarily interested in either making a profit or pushing for promotion. According to sociologist Jonathan Unger[66] this is a problem throughout the PRC because these officials often have little loyalty to the local populace:

> *"Many of [these officials] today put energy during their tenures into showy projects to impress visiting officials [...] in the knowledge that promotions depend upon making a splash. Within a few years they have departed, leaving behind debts that must be paid out of future local revenues."*[67]

Complaints have emerged from Tibetan areas, like those from the rest of China, that officials are busy making quasi-legal financial gains during their tenure. Several accounts accuse officials of taking bribes from mining enterprises.

65 TINDoc 26(jw)

66 Head of the Contemporary China Centre at the Australian National University.

67 Jonathan Unger, 'Power, Patronage and Protest in Rural China', China Briefing 2000: The Continuing Transformation, Tyrene White (Ed), Asia Society/M.E. Sharpe Inc, New York, pp.76-7

Some Tibetans have gone directly to the mine sites to confront workers with their complaints. However, if the miners have official permission to mine, this usually achieves very little. A monk from Ngari (Ch: Ali) prefecture, TAR, said that gold mining being carried out close to his monastery had damaged local pastureland. Nomads living nearby told the workers to stop mining, but the workers simply produced their mining permit and told the nomads to stop complaining.[68] A Tibetan monk from Rebgong reported the same sort of reaction from miners in his area:

> *"There is always trouble [between workers of the mine and the public] if animals die of drinking the water and if animals are wounded because of the explosions. However, the workers say: 'It doesn't help if you cause troubles for us, you have to tell the government. We work for the government'."*[69]

A Tibetan who used to run a small-scale gold mine in Lithang county, Kardze TAP, made clear his attitude as a private miner:

> *"The government gives money to the nomads and the government gives us the right to mine, so the nomads can't tell us anything."*[70]

Community religious leaders can also be outspoken about resource exploitation, both to the authorities and within their communities. For example, a Tibetan with family in Payul (Ch: Baiyu) county, Kardze TAP, said that during the 1990s local lamas were concerned about the impact of small-scale gold mining in the area and had tried to use their influence to dissuade local people from prospecting.[71] However, many of the young people in the area, attracted by the potential wealth to be earned through mining, did not respond to the advice of their religious leaders and continued mining.

The most extreme state response to the opposition to mining, that is known to TIN, was the imprisonment, in 1996. of a senior lama of Nubzur monastery in Serthar (Ch: Seda) county, Kardze TAP. Kabukye Rinpoche was detained on 10 June 1996 after making written representations to officials complaining about the influx of Chinese into the area and the exploitation of natural resources, in particular a gold mine on a mountainside behind Nubzur monastery. He is reported to have raised concerns that blasting operations to loosen rock on the hillside were causing problems for the nomads herding their livestock, and that the mining was leading to the erosion of the grasslands.

68 TINDoc 53b(ke)
69 TINDoc 554(jw)
70 TINDoc 446(jw)
71 This was reported to TIN as concern over the 'environmental damage' caused by gold mining. Although the cause for concern was similar to the western understanding of the environment, ie damage to land and waters, the concerns of the religious leaders are likely to have stemmed, at least in part, from traditional beliefs concerning the disturbance of local deities and spirits. See below: Mining and sacred sites

Gold mine behind Nubzur Monastery, Serthar (Ch: Seda) county **© TIN, 1998**

Kabukye Rinpoche was subsequently sentenced to six years in prison for 'counter-revolutionary splittism' and transferred from Kardze to Ngaba prefectural prison in Maowun county (Ch: Maoxian), also known as Maowun prison.[72] According to an unofficial source he was accused by the authorities of being responsible for putting up pro-independence posters in Serthar during the 40th anniversary of the founding of Serthar county on 24 July 1995. Kabukye Rinpoche reportedly denied any involvement with pro-independence activities, but had written a letter to the authorities protesting against mining near Nubzur monastery, and what he considered to be excessive Chinese migration into the area. The gold mine had attracted over 300 Chinese miners since 1992.[73] Despite assurances from the Sichuan provincial mining department that the local economy would benefit from development of the gold mine, local people felt that neither the monastery nor the local Tibetans had in fact benefited.[74] TIN has received confirmation that Kabukye Rinpoche was released in June 2002 on completion of his sentence, although it is not yet known whether he has been able to return to Nubzur monastery.

72 Several Tibetan political prisoners have been transferred from Kardze TAP to serve sentences at Ngaba Prison. As of summer 2001, there were known to be 14 Tibetan political prisoners in Ngaba Prison, including Kabukye Rinpoche. See "New reports on Ngaba Prison", TIN News Update, 27 April 2001

73 According 1990 census data, the total population of Serthar county was 33,646. Of that population 93 per cent was reported to be Tibetan, with only 2,244 Chinese or 7 per cent (Marshall and Cooke, 1997, op cit, p. 436). Official Chinese population statistics generally under-report the number of Chinese in Tibetan and other "minority nationality" areas.

74 For more details see 'Harsh Punishment for Environmental Protest', TIN News Update, 15 December 1998

Direct action

Sabotage is one of the risks inherent in any mining operation, particularly in regions where there is substantial local opposition. While there have been no reports of violent sabotage of mining operations, such as bombings or destruction of mining equipment,[75] TIN has received reports of local people, frustrated with the official channels, taking direct action to make their complaints heard in two areas of Amdo, with markedly different results.

A county-level gold mine in Dzoige Nyima (Ch: Zhuoge Nima) township of Machu (Ch: Maqu) county in Kanlho (Ch: Gannan) TAP, Gansu, opened in 1991. An associated processing plant started operations in 1994 and aimed by the end of 1997 to produce 500kg (half a tonne) of gold per year.[76] According to a Tibetan from the area, there was some opposition to the mine when it was initially developed, with local nomads confronting mine workers.[77] During the 1990s, there appear to have been discussions between the mine management, local officials and the nomads, and an annual payment was made to the township authorities by the mine. However, there was a general feeling amongst the community that while the mine was clearly doing well[78] and county government officials were benefiting, the township received too little. Instead the nomads were aware of the damage to grasslands and were concerned about the failure of the mining unit to backfill land after excavations. Some of them had also lost livestock that drank from contaminated water storage pits and died.[79] By 1999, some of the younger men in the area decided to take direct action to make their complaints heard. In the spring they started what appears to have been a series of protests, camping out at the mine site and halting production.[80]

75 Radio Free Asia (RFA) reported that two Tibetan monks, Tenzin Khedup [sic] and Thubten Tabkai [sic] were sentenced to life imprisonment in November 2001 in connection with an explosion that injured four Chinese gold miners the previous July. A third Tibetan was sentenced to three years (RFA, 11 December 2001). The explosion, initially reported by the Tibetan Centre for Human Rights and Democracy (TCHRD), an Indian based NGO, was reported to have occurred in a miner's tent in Chamdo prefecture, TAR. According to RFA, 16 local Tibetans were taken into detention for questioning following the incident. TIN has been unable to independently confirm the incident and there is insufficient information available on the motivation behind the explosion to assess whether or not it was an attempt to sabotage mining activities.

76 'Caixiang [Tsewang]: the spirit of Gerke', China Today [China Hoy] Spanish Edition, Volume 38, No.7, July 1997, pp.61-63

77 TINDoc 379(jw)

78 According to China Today, the gold processing plant made 23.16 million yuan (approx. US$2.8m) in profits during the period 12 August 1994 (when it began production) to the end of 1996. The article states that the plant "*featured amongst the most advanced companies; it was awarded prizes by leaders at ministerial, provincial and regional levels*" (China Today, 1997, op cit).

79 Reported compensation payments of 1000 yuan per animal dying from contamination would seem to be an official acknowledgement of responsibility on the part of the mine. Steps were reportedly made to improve the environmental safety standards at the mine following the deaths of animals. For further discussion see Chapter 5.

80 Although accounts vary, there appear to have been at least 20 nomads involved in the protests.

One nomad from the area described the way in which the protest at the mine site was dealt with by the authorities, through a mixture of threat and persuasion. It appears that care was taken to resolve the dispute without further aggravating the situation by involving security personnel. The county authorities, who are reported to rely on income from the mine to pay staff salaries, were perhaps disinclined to draw the attention of higher levels of authority to the complaints over mining:

> *"The higher levels sent lamas to us whom we revered and trusted and then they sent leaders whom we like to talk to us. No policemen came, – but they threatened us many times saying that they would come. Furthermore, many officials from the four main departments of the county government came there, not once, but many times. They held extensive education [sessions] and they sent people who were respected among the public. In that way they twisted here and there and sent people to whom we would listen to speak to us and finally they changed the public's mind. They did not use violence by sending armies to end the protest – they worked out a political solution and they didn't have to use any force."*[81]

To resolve the dispute the nomads were reportedly awarded 200,000 yuan (approx. US$24,000) in compensation and a promise was made to employ people from those local families suffering particular hardship. As a result, at least 20 Tibetans from poor families in the township were employed as daily labourers at the mine during the following two years. However, the greatest benefit to the area as a whole was a substantial increase in the yearly payment that the township government receives from the mine.[82] This money is then used to pay the tax obligations of the township population, meaning that local people no longer have to pay tax and are considerably better off as a result. Heavy tax obligations have been a common cause of complaint among nomads and farmers in recent years. The township is also reported to use part of the income to provide poverty relief to some families. Machu is known as one of the richest Tibetan nomadic areas, partly as a result of the gold mine. The mine also tightened contamination control standards to some extent[83] and nomads receive 1000 yuan (approx.US$120) compensation for each animal that dies as a result of contamination, although it is not clear if this compensation system was in place before the 1999 protest.

81 TINDoc 1017(jw)

82 This was reported by one source to be as much as 1.5 million yuan (approx. US$180,722)

83 The authorities fenced off contaminated water pits to prevent animals drinking from them. See Chapter 5. There are still concerns over contamination of surface and underground water.

The outcome of these protests in Machu is highly unusual, representing by far the most successful instance known to TIN of a local community's complaints about mining being listened to and met with substantial compensation. However, there is still concern amongst Tibetans from the area that, despite the short-term benefits, the long-term cost of damage to land and traditional livelihoods is too high for local people. One nomad summed up this conflict of interests:

> *"When I look at how the poisonous water kills many animals and the way they dig the land I feel we are at a great loss. Then when I think that we don't have to pay tax and how our children are supported then we have benefited."*[84]

Machu (Ch: Maqu) grasslands in autumn **© TIN, 2000**

There has also been a reported instance of direct action taken by nomads and farmers in opposition to gold mining in Rebgong county, Malho TAP. The gold mine is located on land that was formerly used as pasture by semi-nomadic villagers. Opposition to the mine has largely taken the form of spoken and written representations passed up from the village through the different administrative levels. However, a monk from a semi-nomadic family in the township said that in the mid-1990s his village attempted direct action, blocking the road to the mine for one day because they had not received the compensation they understood had been promised to them by the county authorities. The outcome of this protest, and the other representations made by the village, is quite different to the situation in Machu county (see above). The monk described what had happened:

84 TINDoc 1011(jw)

"People protested when the mining started, but because the Chinese consider the water and the soil are theirs they would not listen if people said that mining should not start. So the people said: 'Normally this area is used by the public, now there won't be income from this place, so therefore please cut our taxes.' The community told the village head, the village head told the township government, and the township government told this to the county and prefecture government. They made this request each year, but the [higher levels] didn't listen.

"The government is using this area [for gold mining] but they didn't give us any compensation. First they said they would give 30,000 yuan [approx. US$3,600] to the public. When the money was not given, the public blocked the road, stopping the transportation of mining materials. They blocked the road for one day, and they told the [mine] workers, 'Don't come here again.'

"After this, three people including the village head were detained [temporarily]. They had blocked the road and [stopped] the mining, so it was said that they had damaged the construction of the nation.[85] This is what the Chinese always say. They were detained in the county detention centre and told they couldn't do such things in the future. They were told that this area was not theirs, that they were not the owners of this area, that the land and water belonged to the state.

"People did not block the road again after this, but they would go to the township. They talked many times to the township [government] about this. Letters were given to the township, county and prefecture levels. The letters asked them not to do this to the public, that if they wanted to use the area it would be okay if they gave some compensation because people in the area depend on the pastures."[86]

The above testimony suggests that while local people were actually opposed to the mining, they felt that any request for a complete halt to extraction would be futile, so they requested compensation instead. A nun from Jomda county in Chamdo,

85 According to Chinese law, anyone disrupting the order of mining is open to criminal investigation and prosecution. Mineral Resource Law of the People's Republic of China (passed in 19 March 1986; amended 29 August 1996):
Article 41 Anyone who steals or seizes mineral products or other property of mining enterprises or exploration units, damages mining or exploration facilities, or disrupts order in production and other work in mining areas or in areas of exploration operation shall be investigated for criminal responsibility in accordance with relevant provisions of the Criminal Law; if the circumstances are obviously minor, the offender shall be punished in accordance with relevant provisions of the Penalty Regulations Regarding Public Security Administration.
86 TINDoc 554(jw)

where the Yulong copper mine is located, expressed the lack of power she feels Tibetans have over the exploitation of local resources and the frustration of local people unable to make their voices heard:

> *"Everyone in the valley, whether young or old, worries and complains that the Chinese are taking our minerals. However, our shouts are too small for this big nation. They won't heed our cries, which are like cries of insects to them."*[87]

Political demonstrations

Calls to halt mining can be part of more general dissatisfaction with Chinese policy and have been an element in some political demonstrations. One Tibetan from Qinghai talked about students taking part in a demonstration when he was at middle school in the early 1990s:

> *"The students went to the gate of the prefectural government office. Each class submitted a memorandum to the government office calling for an end to political re-education and an end to mining. Nothing happened. A similar procession had been led in 1989 too. It was also on the same issues."*[88]

A Tibetan from a different area in Qinghai remembers a similar protest when he was at school. His comments show that although the issues that fuelled the students' demonstration were directly linked to the changes taking place around them (including mining), they expressed their concerns through the contemporary political language of the pro-independence movement, even though they did not really understand the concept of a 'Free Tibet':

> *"When we were in middle school in 1987/1988, students protested against the cutting of trees, mining, against sterilisation of women and the fact that there were more and more Chinese and fewer Tibetans. At that time 300 students went to the demonstration. There were 1,000 students in our school but the young ones couldn't go. We shouted 'Free Tibet' although we didn't really understand what this meant – we had been educated by the Chinese."*[89]

87 TINDoc 494(jw)
88 TINDoc 244(jw)
89 TINDoc 6(rp)

The failure of the authorities to allow the average Tibetan a participatory role in development, or to take account of their needs in its process and outcome, inevitably fuels antagonism. However, the authorities aim to promote an image of stability and growing prosperity in Tibet and do so partly by publishing reports on the enthusiastic response of Tibetans to local development, including mining. The idealised picture painted by these reports omits any hint of local opposition and fails to address the difficulties facing Tibetan communities as a result of such development.

The official Tibet Daily report of the opening of the Gyama Trikhang (Ch: Jiamachikang) multi-metals mine in Maldrogongkar (Ch: Mozhugongka) is a typical piece of state propaganda, that promotes the idea that Tibetans unconditionally welcome mining. The mine, which is one of the TAR's main mineral extraction projects,[90] was officially opened on 28 April 1993 and according to Tibet Daily was welcomed by 'the masses':

> *"When Dargye, the chairman of the Lhasa City Economic Planning Commission, announced: 'The Ceremony of Gyama experimental pilot multi-metals mine is to be opened', there was at once an orchestra of the sounds of firecrackers, shells, applause, cheers and songs all over the Gyama valley coming from the celebration meeting, from the masses, and from the loudspeakers."*[91]

However, unofficial reports suggest that there was considerable opposition to the mine amongst locals in Gyama township. Although the mine was officially opened in 1993, associated infrastructure construction had been carried out in the area since the late 1980s, and the impact on people's lands and livelihoods appears to have been one issue fuelling the resentment of Tibetans in the area. General opposition to Chinese rule and policy, including political education and mining, manifested itself in a burst of peaceful protest activity in Gyama during the late 1980s and early 1990s. Some of the political posters that appeared in the area between 1989 and 1992 expressed opposition to mining and unconfirmed reports indicate that three Tibetans from the area were detained in March 1992 for putting up such posters. One of these Tibetans, 18 year old Phuntsog Choesang, was reportedly held for 14 months without legal process in Maldrogongkar county Public Security Detention Centre.[92]

90 See Chapter 1, p.34

91 Tibet Daily, April 1993 in "Bod kyi pho nya", vol.1 1993, Guchusum.

92 The other two Tibetans, Gyatso and Karma, were also reportedly held at the county PSB detention centre for an unknown period of time. Three months later, in June 1992, four farmers were arrested after they interrupted a political meeting in the township by walking on to the platform waving a Tibetan flag and shouting pro-independence slogans. They were charged with instigating "counterrevolutionary propaganda" and received 13 to 15 year sentences. One, Sonam Rinchen, sentenced to 15 years, died in custody under unknown circumstances (see "Tibetan farmer serving 15 years for peaceful protest dies", TIN News Update, 12 April 2000). A second, Konchog Lodroe, sentenced to 13 years, was released early on medical parole. The other two farmers, Sonam Dorje and Lhundrub Dorje are currently serving their respective 13 and 15 year sentences in the TAR Prison Number One (known as Drapchi prison) and are due for release in 2005 and 2007. A fifth Gyama farmer, Thubten Yeshe, was arrested a few days later. He currently serving a 15 year sentence in Drapchi prison and is due for release in July 2007.

Mining and sacred sites

> *"There is scarcely a peak in Tibet which would not be regarded as the abode of a mountain-god or goddess."* Tibetologist Réne De Nebesky-Wojkowitz[93]

Although many of the concerns that Tibetans have expressed over mining have been pragmatic and/or political, opposition to mining also arises from traditional religious and cultural beliefs and the relationship of Tibetans to their lands and waters. According to popular belief, the mountains, rocks, waters, forests, plains and soil of the Tibetan plateau are inhabited by various spirits and protective deities who can be malevolent if they are not propitiated or if they are offended. Mining activities by their nature disturb the homes of these spirits and deities and as a result are often blamed for any negative occurrences. For example, in addition to their practical concerns over the gold mine behind Nubzur monastery, Tibetans in the area were also concerned that the unexpected deaths in February 1996 of the Abbot of Nubzur monastery and another senior lama were linked to the disturbance of local earth spirits through mining.

One of the most sacred of sites in Tibet is Mt. Kailash and its two adjoining lakes, Manasarovar and Rakas Tal, venerated by Buddhists, Bonpo and Hindus. According to Tibetan historical accounts, gold mining was carried out at Lake Manasarovar (Tib: Mapham Tso) in 1990, but was stopped by the Tibetan government following an outbreak of small pox that was attributed to the wrath of the presiding deity of the region.[94] Gold exploration and mining activities had recommenced in the area during the 1990s and continue at present (July 2002). By the end of the 1990s, there were reported to be about 100 workers at one mining site, located on the west side of the lake near to Ji'u (Wylie: Byi'u) monastery. The miners had built an impoundment on the Ganga Chu, the channel connecting Manasarovar to Rakas Tal (Tib: Lhanak Tso).[95] Water was being pumped from the impoundment to provide water for exploratory prospecting about 3km south of the channel. Power for the operation came from a fossil-fuel operated generator located near the mine camp.

93 Réne De Nebesky-Wojkowitz, 'Oracles and Demons of Tibet: The Cult and Iconography of the Tibetan Protective Deities', Book Faith India, 1996 (originally published 1956).

94 'Tibet 2000: Environment and development issues', Environment and Development Desk, Department of Information and International Relations (DIIR), Tibetan Government-in-Exile, Dharamsala, 2000

95 According to legend, the Ganga Chu was created when a golden fish from Manasarovar tunnelled through the earth to let sacred water through to Rakas Tal in order to purify the poisonous waters of the latter. Rakas Tal is named after the flesh-eating demons of Hindu mythology believed to lurk within its waters. (Victor Chan, 'Tibet Handbook: a pilgrimage guide', California, Moon publications, 1994, p.614).

Gold mining at Lake Manasarovar: electricity lines and water hoses above the impoundment on the Ganga Chu © TIN

Gold mining at Lake Manasarovar: Impoundment on the Ganga Chu © TIN

According to Tibetologist René De Nebesky-Wojkowitz, lakes in Tibet are frequently regarded as being *bla gnas* – the place to which a man's, or even a nation's, 'life-power' (Tib: *bla*) is attached.[96] Yamdrok Tso is one such lake and, the legend goes, if it should run dry the whole population of Tibet will die.[97] Similarly, author Victor Chan, who has travelled extensively in Tibet, says that the level of the Ganga Chu is associated with the state of the country:

"High water augurs well for Tibet and its citizens. During much of this [20th] century, the salt-encrusted channel was virtually dry [...]. Only in the last few years has water flowed between the two lakes again; some say this coincides with the resurgence of religious practice in Tibet [following the Cultural Revolution]".[98]

There are also clearly environmental concerns connected to mining in this district; four of Asia's major rivers have their sources in the immediate area: the Indus, the Sutlej, the Tsangpo (Brahmaputra) and the Karnali, while the Ganges originates close by.

One case reported to TIN illustrates the attempts that are sometimes made by mining enterprises to placate those with religious/cultural concerns about mineral exploitation. A Tibetan from Lhoka (Ch: Shannan) prefecture told TIN about a mining operation in the area of a holy mountain. The only monastery in the area made a complaint about the mine.

"When the Chinese mining was in progress, the monastery argued and gave its reasons why it wasn't allowed. At that, the mining commission talked of building roads for the monastery and agreed to solve other such problems the monastery was facing. They also organised a grand chanting service at their expense".[99]

Tension between traditional beliefs and mineral exploitation is not unique to Tibet. Mining companies have had to face the issue of potential violation of sacred sites in many countries. The issue becomes particularly sensitive when mining is being carried out on indigenous lands by a politically and economically dominant group with differing religious and cultural beliefs and different priorities. It is precisely these kinds of traditional Tibetan beliefs in the need to pacify the spirits and deities of rivers, lakes and mountains that the atheist Chinese state regards as the 'feudal and superstitious dross'[100] holding Tibet back from development. However, many Tibetans see no tangible benefits to mining and instead have to live with the damage to local land and waters. This may not only spark opposition to mining, but would also appear to validate traditional beliefs prohibiting the disturbance of earth, rocks and waters.

96 Réne De Nebesky-Wojkowitz, op cit.

97 Ibid. The authorities have constructed a 90 mW pumped-storage hydropower plant on Yamdrok Tso to power Lhasa.

98 Victor Chan, 1994, op cit.

99 TINDoc 22(vn)

100 For example a commentary on Tibet TV in August 1999, used the *Falun Gong* issue to highlight the dangers of "idealist theism" in Tibet and to promote atheism. The commentary asked the question: "*Do we dare in light of Tibet's reality to squarely face the harm of religious idealist theism and its feudal and superstitious dross to the building of material and spiritual civilisations in Tibet?*" ('TIN News Review: Reports from Tibet 1999', London: TIN, 2000).

Gold mining at Lake Manasarovar late 1990s, clockwise from top left: trench running south to mining camp; digger extending trench; water pours into holding tank; sluicing machinery (top right of image), not yet in use © TIN

4. Labour issues

International best-practice standards for the mining industry support the principal that local people should benefit from mining in their localities by gaining access to employment. However, it is a common problem that local people rarely have the skills needed to engage in the better-paid jobs that have greater responsibility and a higher status. As a result skilled workers from elsewhere generally receive the best opportunities and existing economic and social disparities are often enhanced to the detriment of local people. Those who argue for mining to contribute to sustainable development stress the need for local training and the development of alternative opportunities that will prevent the marginalisation of indigenous inhabitants and sustain local economic development beyond the life of a mine.[1] At present, most Tibetans are nomads and herders and do not have the education or training to find much work beyond basic unskilled labour that lasts only as long as a mine is in operation. Current development policy indicates that this pattern is likely to be perpetuated with the drive to develop the western regions of China and expand Tibet's mining industry.

Although employment in the mining industry can be viewed as a relatively good opportunity, it can also be very risky, depending on living and working conditions. Injury, illness and even death as a result of exposure to dangerous chemicals or accidents caused by poor engineering, for example tunnel collapses, are among the risks that mine employees can face. Conditions in China's mines are among the worst in the world, with general health and safety standards far below basic international benchmarks. Workers rights are not protected and independent trade unions are not permitted. In addition, the use of forced labour persists throughout China as the basis of administrative and judicial penal systems, and through the conscription of local communities in economic construction.

The first section of this chapter looks at the involvement of Tibetans in the mining industry, which has largely been limited to manual labour and artisanal gold mining, noting that while the predominance of Chinese in the workforce varies from area to area, the majority of skilled and managerial positions appear to be held by Chinese employees. The reasons that Tibetans remain disadvantaged in employment in the mining industry are considered, including the need for education and skills and the important role that '*guanxi*' (connections) has played for those Tibetans who have found skilled labour or more senior positions.

1 For example, see: 'Breaking new ground: mining, minerals and sustainable development', MMSD (IIED), 2002. The full text of the report can be downloaded from the IIED website www.iied.org

The second section discusses pay and conditions for Tibetans working in the mining industry, considering that Tibetan workers face many of the same problems as industrial employees throughout China, including poor health and safety standards that do not appear to be any better than the notoriously low standards throughout China. The last section of this chapter deals with the issue of compulsory labour, charting the use of prison labour in the development of Tibet's mining industry and exploring the difficulties in establishing the extent to which community labour used to support economic development, including mining, can be considered voluntary.

Access to employment

The majority of Tibetans are nomads and farmers. The work that they are likely to find in the official mining industry is generally limited to low-wage unskilled labour, such as breaking up rocks, digging and loading trucks or machines, or providing manual labour for infrastructure construction (roads, houses, power). However, even in the most basic jobs Tibetans are increasingly facing competition from Chinese migrant workers. Some Tibetans have found work in mines as a 'sideline' occupation, casual seasonal work[2] to supplement low income from animal husbandry or farming. Others have been engaged in private mining ventures and artisanal mining. The majority of skilled and managerial positions appear to be held by Chinese employees. This is unlikely to change in the near future. Tibetans remain disadvantaged in language, education and skills and although the Chinese government states that it supports the improvement of vocational education for Tibetans, it is primarily concerned with immediate, accelerated development. As a result the authorities are encouraging the importation of personnel from central and eastern China to fill skilled positions that are opening up as the industry develops. Despite the increased focus in China on granting employment on the basis of competency, employment opportunities for Tibetans, as for workers throughout China, remain heavily influenced by their '*guanxi*' (connections).

2 Due to the Tibetan climate mines are generally unable to operate during the winter months.

Manual labour

The dominance of Chinese in the unskilled workforce varies from area to area. The mines that were established in Qinghai during the 1950s continue to rely on the labour of large numbers of Chinese settlers, many of whose families originally came to the province in the 1950s and 1960s as forced immigrants or prisoners. Over the last 20 years, the opening of the Xining-Golmud railway and policies to encourage voluntary immigration have provided a flow of migrant labour for the accelerated development of the mining industry. The mining towns of the Tsaidam Basin in Tsonub (Ch: Haixi) Mongolian and Tibetan Autonomous Prefecture (M&TAP) were established in sparsely populated areas that, before the 1950s, were inhabited by Tibetan and Mongolian herders. The predominantly Chinese population of these towns today reflects the history of the mining industry in Tsonub, which was established and developed largely by Chinese settlers.[3] Tibetans and Mongolians have generally remained on the fringes of this economic development.

By contrast, the Norbusa (Ch: Luobusa) Chromite Mine located in Chusum (Ch: Qusong) county, Lhoka (Ch: Shannan) prefecture, Tibet Autonomous Region (TAR), apparently has a largely Tibetan workforce. Built on the site of the Dongfeng iron mine (opened up in 1967), Norbusa is the largest chromite producer in China and one of the TAR's key mining projects. The director of the mine is Chinese, but according to the mine's own publicity material (undated), of the 624 mine employees, 576 were Tibetan.[4] The high rate of Tibetan employment at the mine may reflect the demographic difference between Qinghai and the TAR; the population of the latter generally remains predominantly Tibetan, with the notable exception of Lhasa municipality.[5]

However, the official predominance of Tibetans in a local population is by no means always reflected in the numbers of Chinese and Tibetans employed in manual labour in larger mining enterprises. According to a former government official, Chinese workers were being brought into Tsigorthang (Ch: Xinghai) county Tsolho (Ch: Hainan) Tibetan Autonomous Prefecture (TAP) in summer 2001 to prepare infrastructure (power, roads, housing) for the Saishitang copper deposit. At this stage, locals were not being offered employment in the mine.

3 For example, the combined populations of Mangya, Lenghu and Dachaidan were 95 per cent Chinese according to 1990 census data. See Chapter 3 for further details of immigration, and specifically a study of the demographic development of the Tsaidam Basin and its relation to the mining industry.

4 'Rhobusa of Tibet – the largest Chromite Mine in China', published by the Norbusa mining office in English, Chinese and Tibetan, undated.

5 According to official population data the TAR remains 96 per cent Tibetan (TAR Statistical Yearbook, 2001; the TAR 2001 yearbook does not include 2000 census data). Although the numbers of Chinese are understated (the military and the floating population are not included), the region still retains a much higher proportion of Tibetans than Qinghai province. According to 2000 census data, Qinghai is only 22 per cent Tibetan. Even if one only takes into account the total population of the six prefectures with Tibetan autonomous status, the Tibetan population still only represents 53 per cent. (Qinghai Statistical Yearbook, 2001).

According to 2000 census data, the county was still 77 per cent Tibetan.[6] The former official said that the workers were living in tents, but housing was being built for them and that thousands more Chinese workers were expected. Similarly, employees at the Gyama Trikhang multi-metals mine, one of the TAR's key mining projects, are reported to be predominantly Chinese. Tibetans have been employed in road construction for the mine, although there have been conflicting reports as to whether they were paid for this work.[7]

According to a Tibetan from Jomda (Ch: Jiangda) county in Chamdo (Ch: Changdu) prefecture, TAR, a few local Tibetans have found employment at the Yulong copper mine as guards, while others are reported to have been employed as daily labourers in road construction. The surveying teams that have visited the area have been Chinese and they have brought Chinese workers with them and sometimes Tibetans from other areas. It is not yet clear what opportunity there will be for Tibetans to find manual work at the Yulong mine once it is fully operational.

Tibetans may have more success finding work in smaller-scale ventures than in new larger-scale state projects, where workers often appear to be contracted and brought in from outside the local area. Tibetans in some areas have also engaged in artisanal gold mining. However, in both cases they still have to compete with migrant Chinese labourers, particularly in those areas near to densely populated Chinese regions with thousands of under-employed rural workers.

For example, during the 1980s and 1990s, there was a gold production drive in Kardze (Ch: Ganzi) TAP, Sichuan, initially focusing on the development of collective enterprises. During the 1990s, private gold mining in Kardze prefecture was also encouraged by local authorities who would lease out land or issue mining permits to prospectors. In most areas the mining sites were predominantly worked by Chinese, but local Tibetans were also involved in prospecting. Steven D. Marshall and Susette Cooke, who carried out extensive field work in the area in the mid-1990s, commented that in some areas:

> *"At times it appeared as if entire villages turned out for a communal effort at digging and sieving".*[8]

6 Qinghai Statistical Yearbook, 2001. See Chapter 1, pp. 33-34 for further details of the Saishitang copper mine.

7 See below, p. 143

8 Steven D. Marshall and Susette Cooke, 'Tibet Outside the TAR', CD-ROM published by the authors, 1997. There have been unconfirmed reports of Tibetans in Kardze prefecture being forced to engage in gold mining by local authorities (see section on compulsory labour below).

However, Marshall and Cooke also noted that in most areas, prospectors were predominantly Chinese. Exclusively Tibetan sites were often much smaller and less organised than Chinese sites. In some areas Tibetans appeared to be working Chinese mines after the sites had been abandoned by Chinese gold miners with more advanced equipment.[9]

The account of a monk from Pari (Ch: Tianzhu) Tibetan Autonomous County (TAC), Wuwei prefecture, Gansu, interviewed in exile in 1995, suggests that a similar practice had been occurring in his area, where the government would mine land and then lease it out to private individuals when they were finished:

> *"There is a lot of gold mining in the area. Workers are local Chinese. They use track vehicles driven by military personnel. There are many places where they search gold, all depending on the goldmine office. They work on a spot for one or two years, digging the ground and back-filling it when the search is complete. When this is done, they hire the land out to individuals: it costs 30,000 yuan (US$3614) for about four square metres. Both Chinese and Tibetans are searching for gold in this way. They are closely watched by the police as they have to sell their product to the goldmine office. The rate of the black market is 1.5 times higher than the government rate."*[10]

A Tibetan from Lithang (Ch: Litang), Kardze TAP, talked about small-scale private mining in his area. He said that the government started leasing out land to people from the area in 1995, but then from 1997 opened up the business to people from outside the county. He says that thousands of Chinese people arrived in Lithang. According to his reckoning there were about 100 sites being worked, each employing an average of 100 Chinese labourers, the number depending on the size of the site being leased. He himself secured a contract to mine an area of land and employed Chinese workers:

9 Marshall and Cooke, 1997, op cit.

10 TINDoc R40(RM2). Despite Pari's Tibetan autonomous status, Tibetans only accounted for 26 per cent of the county's population in 1994 (Marshall and Cooke, 1997, op cit). Pari TAC has a relatively developed mining industry. According to Marshall and Cooke, official sources indicate that there are numerous mining sites throughout the county, extracting coal, quartz, gypsum, iron, copper, lime, nickel, titanium and other minerals (Annals of Tianzhu county, map). Uranium deposits are also found within the county (Tibet – A Land of Snows). In the mid-1980's coal mines in Pari were producing over 200,000 tonnes of coal per year (Tibet – A Land of Snows). See Marshall and Cooke, 1997, op cit.

"When I first started mining I employed 90 Chinese workers. We mined for four months and we didn't make any profit. Then I fired 30 of the workers and kept 60. One of my relatives does business in China. He brings Chinese people he knows with him [to Lithang]. Later, if they don't have any work anymore they stay and collect Yartsa Gumbu [caterpillar fungus].[11]

He explained that due to the costs of leasing the land, employing labourers and other overheads (including the necessity to bribe local officials), most Tibetans are not in a position to start their own business in this way:

"About 20 per cent [of people able to buy land] are Tibetan. The others are Chinese from Sichuan, Qinghai, Gansu and Yunnan. Most Tibetans don't have enough money to start gold mining like this."[12]

Many Tibetans involved in the mining industry are engaged in gold mining as a 'sideline' occupation, a way to supplement their income from animal husbandry or farming by getting seasonal work in mines. A Tibetan farmer from Haidong (Tib: Tsoshar) prefecture in eastern Qinghai said that people in his village had to find alternative sources of income in order to survive. Tsoshar is the most densely populated of Qinghai's prefectures, comprising only two per cent of the province's land area, but over a third of its population. Population pressure and land degradation have become acute through development policies that in the early years of the People's Republic of China (PRC) included forced immigration.[13] For this Tibetan, work in gold mines was one option for people in his village to supplement their low incomes:

"We don't have enough food from our crops. For that reason, people from our village go to do side jobs. People do whatever they can do. People do work like road construction, construction of buildings and mining of gold. After planting the crops in the spring most of the people leave our village to do side jobs."[14]

A Tibetan who worked at a coal mine in Wulan county, Tsonub M&TAP, said that most of the Tibetans working at the mine were migrant labourers from the north-east of the province.

11 TINDoc 446(jw)

12 Ibid. It is useful throughout China to have good connections (*guanxi*) to access funds/loans to set up businesses and to obtain relevant permits. The use of connections and bribery that this source employed to start his mining operation in Lithang is mentioned below, p. 129

13 See Susette Cooke, 'The Politics of Population Transfer', TIN Special Report, 28 October, 1999. According to 2000 census data nine per cent of Haidong's population is Tibetan. See note 76

14 TINDoc 343(jw)

> *"There were only a few local people working at the coal mine. A few were driving vehicles, a few were working with the explosives and there were about two of them that were working with the machines. There were more than 300 workers that dig coal and shovel it into the vehicles. There were both Tibetans and Chinese, but not Tibetans from our area, these people looked like they were from Hualong [in Tsoshar prefecture]. They were paid daily. If two people shovelled coals in a vehicle I think they were paid about 15 yuan (US$1.8)."*[15]

Tibetans with driving licenses and their own truck have been able to pick up casual seasonal work in some areas, but opportunities tend to come and go. A Tibetan who worked as a driver at a gold mine in Machu (Ch: Maqu) Kanlho (Ch: Gannan) TAP, Gansu province in 1997 talked about the work:

> *"I worked there for two months, transporting the rocky stone from the pit to the engines where gold is extracted. My truck is my own private truck, they paid me to work there. For one trip a truck can carry between 12 and 14 tonnes of stones and we were paid 10 yuan (US$1.2) per tonne of stones. We weren't paid daily or monthly, but we were paid after we completed the work [at the end of the season]."*[16]

A nomad from the same area said that this opportunity for casual labour had been lost in the further development of the mine; by 2001 the mine exclusively used its own vehicles and drivers:

> *"The vehicles of the gold mine now tow away the rocks. We had already arranged to take turns in towing the stones but they said that from this year they have many vehicles of their own and so they don't allow us to do that work."*[17]

Both casual labour and artisanal gold mining have constituted a form of poverty relief for poor Tibetans and Chinese. In recent years China has attempted to increase central (macro) control over mining and to curb semi-legal and illegal mining. According to a Xinhua article published in February 2002, there were still 20,000 illegal prospectors entering Qinghai's gold fields each year.[18] Alluvial gold mining was officially banned throughout Qinghai province in February 2002, with the exception of three state mines. Many of the prospectors were farmers living in the overcrowded,

15 TINDoc 643(jw)
16 TINDoc 67(jw)A1
17 TINDoc 1017(jw)
18 'Qinghai Province Bans Gold Digging to Protect the Environment', Qinghai Daily, 9 February 2002

predominantly Chinese, north-eastern part of the province who engaged in gold mining as a form of poverty alleviation. An article in the Xihai Metropolitan News acknowledged the economic impact of the gold mining on tens of thousands of farmers in Kumbum (Ch: Huangzhong) county, Xining municipality, stating that:

> *"For many peasants, their only source of income to support their families apart from farming is digging for gold".*[19]

Chinese gold miners eat breakfast, north of Dege county town, Kardze (Ch: Ganzi) TAP © TOTAR, 1997

The reports adds that this labour has now been channelled into road building, urban construction and the Golmud-Lhasa railway. The majority of these farmers are Chinese or Hui – according to 2000 census data Tibetans only account for about eight per cent of the county's population. However, the impact of any crackdown on small-scale mining will also be felt by those Tibetans who have engaged in mining as 'sideline' job, either as casual labourers or as artisanal gold miners.

In Ngari (Ch: Ali) prefecture, TAR, the authorities appear to be channelling surplus rural labour into the mining industry. According to a Xinhua report date 11 June 2002, Ngari plans to transfer excess labour in Gergye (Ch: Geji) county, in total 862 people, to mining areas and building sites.[20]

19 '200,000 Kumbum Peasants Leave to Earn Money', Xihai Metropolitan News, 9 May 2002. NB: The number of peasants that have left the county varies in the article from between 40,000 to 200,000.
20 'Gergye county extends scope of labour transferral', Xinhua, 11 June 2002, on www.tibetinfor.com

Labour as compensation

There have been a few reports of mines or associated processing factories offering employment to locals as a form of compensation for mining or other construction that has adversely affected the livelihoods of nomads and farmers.

Some Tibetans in Machu from poorer families were given work at the gold mine there following local protests.[21] According to a nomad from the area they are employed as 'daily workers', rather than salaried staff, and are engaged in manual labour, for example breaking rocks or guarding the contaminated water storage pools to keep animals away from them. One of his relatives was employed a daily worker, but was later offered regular work in the mine:

> *"We are a poor family and one of my relatives was given the chance to work there. At first [my relative] was separating the rocks containing gold from the rocks not containing gold after the ore was taken from the mine. Now he has been promoted and is a leader of the section. He has to supervise the work of separating the gold from rocks. These days he is a permanent employee and his salary has increased."*[22]

According to an aluminium factory worker from Rebgong (Ch: Tongren) in Malho (Ch: Huangnan) TAP, work in the mining industry there was offered as part of a compensation package to farmers whose farmland had been requisitioned for state construction. In addition to monetary compensation for land requisition, one child from each family was allowed to work in the aluminium factory or nearby gold factory.[23]

The possibility of employment in the mining industry for disadvantaged local people is by no means universal. For example, the former government official who reported the arrival of Chinese workers in Tsigorthang county to develop the Saishitang copper deposit[24] said that, although the county had requested that unemployed people from the township and county would be employed in the mine, so far no one in the township had been given a job.[25]

21 see pp. 105-107
22 TINDoc 1011(jw)
23 TINDoc 746(jw)
24 See above, p. 117
25 TINDoc 8(rp)

Skills/education

Work as a semi-skilled or skilled labourer (operating machinery for example) and in the management of mining enterprises can be considered by Tibetans to be good employment where other opportunities are lacking. However, it still appears that Chinese workers take many of the jobs requiring technical skills or managerial responsibility. Technical and scientific jobs requiring higher education (for example in exploration geology) appear to be held almost exclusively by Chinese. This reflects the general trend in most sectors of urban or industrial employment in Tibet. Faced with the continued immigration of Chinese workers, Tibetans are finding it increasingly difficult to compete in the job market.

The Chinese authorities have started calling for employment on the basis of skills, something new in Communist China, where ideology was originally the principal determinant in suitability for work, followed by '*guanxi*' (connections) as corruption became increasingly evident throughout the country in the 1980s. However, the majority of Tibetans remain disadvantaged in terms of language, education and skills. In a predominantly Chinese industrial and business environment, Chinese language skills are becoming increasingly essential for Tibetans if they want to compete for jobs. A high level of Chinese language is also essential for those Tibetans who wish to pursue a scientific career (for example to become a geologist). There are very few opportunities for scientific study in the Tibetan language. However, there is also understandable resistance to Chinese language-based learning amongst some Tibetans, due to a desire to protect and develop Tibetan language and culture in an increasingly Sinicised environment. This places both Chinese authorities and Tibetans in a 'Catch 22', no-win situation.

The government has said that it will give support to improvements in vocational education and training in Tibetan and other 'minority nationality' regions of China, to be funded by commercial bank loans and foreign aid.[26] This will be essential if Tibetans are to be able to engage in economic development. But the authorities, committed to accelerated development, are offering state subsidies for the importation of skilled workers from the rest of China.[27] In the early years of the PRC there was a policy of population transfer to support development in some Tibetan areas, most notably Qinghai province. The emphasis has since shifted to the 'poor quality' of the population of 'minority nationality' areas, and the authorities have been calling for a

26 'Implementation Opinions Concerning Measures Pertaining to the Development of the Western Region', State Council office of Western Region Development, 28 August 2001. See Chapter 2, pp. 50-53

27 The authorities have introduced various measures to encourage the westward migration of 'qualified personnel' and to improve the 'population quality' in Tibet and other western regions. See Chapter 2, pp. 50-53

transfer of talent in order to succeed in China's drive to develop the western regions of China. By the time Tibetans have any chance to 'catch up', they are likely to represent an even more marginalised sector of the workforce than at present.

However, vocational training for Tibetans is starting to develop in some Tibetan areas. For example, the Kanlho Tibetan Nationality Comprehensive Vocational School in Tsoe (Ch: Hezuo), capital of Kanlho TAP in Gansu, started a class in mine surveying in 1998. According to a Tibetan who had studied tourism at the school, which was originally a teacher training college, other subjects taught there were electrical engineering and motor mechanics. This student said that a mine surveying class was offered at the school in 1998 and 1999. This appears to have been directly linked to the development of gold mining in the prefecture. The Tibetan student said:

> *"If they are good in their studies they will be sent to work at one of the mining sites in Kanlho. If they are not good in studies they won't get any work. When they were doing their studies they were taken to Machu to practise geological surveying."*[28]

There has also been some scope for Tibetans to receive training in the work place, either through formal programmes or on their own initiative. Official reports on 'model' mines, indicate that, at least officially, the employment and training of Tibetans is recognised as important. In a China Daily article about the gold processing factory affiliated to the gold mine in Machu, reportedly pioneered by a Tibetan, emphasis is laid on equality of employment and the benefits of training and education:

> *"The plant is a place inhabited mainly by Tibetans and the employees and workers come from three investment companies in which Tibetans also comprise more than half. [...] In order to improve the technical training of the employees, Tsewang sent more than 30 people to the Institute of Gold Research in Shenyang and to other mining regions, where they received specialised training. Taking note of the fact that the majority of the young workers were from agricultural and ranching areas with low education levels, he invited the technicians to give them lessons. Now all the employees and workers, 196 in total, have been trained in turns; 40 have received university diplomas; and 31 the medium level qualification of the trade."*[29]

28 TINDoc 1011(jw)

29 Caixiang [Tib: Tsewang]: the spirit of Gerke', China Today [China Hoy] Spanish Edition, Volume 38, No.7, July 1997, pp.61-63

A 1994 Xinhua article on the Norbusa Chromite mine in Lhoka, TAR, also boasted that "*Tibetan miners have been given the opportunity to be trained at colleges in neighbouring provinces*".[30] Unofficial reports are not always as enthusiastic as the official press about the training opportunities offered within mining enterprises. A gold factory worker commented:

> *"Initially those who entered the factory were given three months training in Xining. But when I entered the factory there was no such training. The factory does not have any job that requires education. One has to work with the stones and mud."*[31]

A Tibetan nomad from Golog (Ch: Guluo) TAP, who had worked in an alluvial gold mine in Qinghai during the mid-1990s, said that although most skilled positions were filled by Chinese, in his experience work at the mine was distributed on the basis of individual ability. Although he was not provided with formal training, he managed to gain promotion through teaching himself the necessary skills:

> *"The educated and skilled employees do 'soft works' like controlling engines and machines. 'Hard works' are digging out and purifying the gold ores and transporting boxes [of ore]. At the beginning I was working with a spade on the extraction site, but slowly I was promoted. While I was doing 'hard works' I had learnt about the engines from the people working with them. The [soft] work I did was controlling the engines that separated the gold ore from the sands after the ore had been extracted and controlling the engines to equalise the water supply for the gold mining. I sat on my chair in the engine rooms and if there was anything needed for the engines and its metres, I just had to switch the engines on or off."*[32]

He said that he worked a five-day week, eight hours per day and was paid about 500 yuan (US$60) per month, although this varied and he could get as little as 200-300 yuan depending on how well the mine was doing. He said that he first got a job (manual labour) at the gold mine through competitive exams, although these appear to have been a general test of his Chinese language ability and his civic and political education level, rather than being a test of vocational skills directly related to the job.[33]

30 'Tibetan herdsmen who dig the new China', Xinhua, 29 October 1994
31 TINDoc 539(jw)
32 TINDoc 47(jw)
33 According to the source: "*There were about 120 people sitting the exam. From 120 people, only 20 or 25 were selected for the job. [The exams were] Chinese language and political science in Chinese. The questions in the exam on political science were, for instance: 'What is law?', 'What is a nation?', 'What are the major types of law of the PRC'.*" TINDoc 47(jw)

Connections

Most of the Tibetans who are reported to have been employed in skilled labour or managerial positions in the mining industry have had good '*guanxi*' (connections), usually a relative working in the local government or Party bureaucracy.[34] Despite the growing emphasis in China on skills-based employment, personal connections can still be more important than education or skills and sometimes they even help people to find employment in manual labour.

A Tibetan who worked at a mineral processing factory in Qinghai described the procedure for applying for a job:

> *"When I entered the factory I first had to write an application to the human resources office and after they had given permission I had to write an application to the manager/contractor of the factory."*[35]

While this procedure seems to be straightforward, he made it clear that he only got the job because a relative of his who had a senior position in the local government arranged it for him through his contacts:

> *"Except for children of high officials it is very difficult to join the factory. My relative got me the job. You cannot find a [good] job without a connection. These days, even if someone joins the military he doesn't get a job automatically. One must bribe to get a job."*[36]

In this case, his relative did more than simply use his connections at the factory. In order to get the job this Tibetan was required to show his middle-school graduation certificate – however, he had never finished school. His relative used his friendship with the headmaster of the local middle-school (a friendship in which, as an official, he was probably the more powerful party) to arrange for a false graduation certificate to be made. This Tibetan worked on the factory floor for two years, operating stone grinding machinery. His relative's position had not only helped him to get the job, but

34 It is important to acknowledge here that status and connections can be significant factors for people seeking advantage or advancement in employment throughout the world. However, the widespread use of '*guanxi*' in practically every walk of life throughout China, and associated corruption, has been noted as one of the threats to China's continued growth and development. As a result the authorities are starting to focus more on the need to employ properly 'qualified' personnel. According to Article 7 of the International Covenant on Economic, Social and Cultural Rights (ICESCR), which China has ratified: 'The States Parties to the present Covenant recognise the right of everyone to the enjoyment of just and favourable conditions of work which ensure, in particular: [...] (c) Equal opportunity for everyone to be promoted in his employment to an appropriate higher level, subject to no considerations other than those of seniority and competence.'

35 TINDoc 9(rp)

36 Ibid.

also enabled him to keep it. He said that when he was going to be fired because he had been involved in several fights in the factory, his relative used his influence to ensure that he kept his position.

PLA base converted into gold-mining support facility, Lithang (Ch: Litang). A slogan inside reads 'Gold is truly beneficial; gold is truly life'. **© TOTAR, 1997**

A nomad from Machu knew a few people who worked at the county gold mine. His account offers a contrast between the different opportunities offered to Tibetans. One of his friends got a job at the mine in a semi-skilled position because 'the brother of his father had a high position':

> *"I have been to the mining site because a friend of mine is working there. He is in control of the water supply of the extraction procedures. His work is fairly light, he has a wireless phone and the only thing he has to do is stop or start the water supply when he is told to do so by phone. So the work is not heavy, but it is a job with responsibility. He is for example responsible if there is a theft. His salary is more than 1000 yuan [per month] (approx US$120) and he has a big house in the mining compound itself."*[37]

37 TINDoc 379(jw)

The same Tibetan also knew two other families who worked in the mine, living there in tents, and employed in manual labour. In talking about the kind of work that these families did, the nomad comments on the general division of labour between Chinese and Tibetans:

> *"They [the Tibetan families] worked in carrying out the ore from the digging sites in wheel-barrows. They get paid five yuan (US$0.6) per wheel-barrow load. Sometimes they have to push these loads a long way, sometimes not so far. It is mostly Tibetans who do this work. They come from all around. There are also Chinese who do this work, but most of the Chinese do work which involves responsibilities."*[38]

Although employment on the basis of connections can have little to do with competence, it is sometimes simply the case that the children of officials (similar to many Chinese competitors for jobs) will be more likely to have received a better education and are therefore better qualified. This reflects the more general international norm that greater power, status and wealth can generate better opportunities for the next generation. In Tibetan areas, as throughout China, the children of Party and government officials are likely to receive a better education than, for example, the children of farmers who have no 'backdoor' channels to gain preferential treatment. For example, a worker at the Rebgong aluminium factory said that the head of the factory is a Chinese from Lanzhou, but that of the three managers below him, two are Tibetans. One, he says, was the son of the county head, the other was the son of a previous county head. Both of them were well connected and both had 'enjoyed very good education'.[39]

Good '*guanxi*' is generally considered throughout China to be an important factor in starting up a private business enterprise – as is the need to bribe the right people. A Tibetan from Lithang talked about his understanding of the way the system works from his experience in setting up his own gold mining enterprise:

> *"I bought land worth 20,000 yuan (US$2416). I gave 6000 yuan (US$724) as a bribe. If you pay well you get a good place [to mine]. There is someone in the office who decides on land and you have to give the money to this person. It is no use giving it to other people – they will just eat that [bribe] themselves. If you don't pay bribes it is difficult to get land [to mine], and even if you do get land they will make it very expensive and will create continuous problems, saying you have dug out too large a hole and so on."*[40]

38 Ibid.
39 TINDoc 746(jw)
40 TINDoc 446(jw)

Pay and Conditions

Wages and benefits

There have been reports of discrimination in wages in the mining industry in Tibetan areas, with Tibetans being paid less than Chinese workers. Based on the limited information available, it appears that this discrepancy in pay is partly due to local authorities using local community/compulsory labour which is generally remunerated at a low wage if at all (see below); or it simply arises from the fact that the work that Tibetans are able to get in the mining industry is generally unskilled labour at the lowest end of the wage scale, as illustrated above. In addition, it has been common for Chinese taking jobs in 'hardship areas' like Tibet to be paid higher wages and receive other perks as incentives.[41]

Some of the reports that TIN has received from Tibetans working in the mining industry reflect the general situation for industrial workers throughout China. The privatisation and restructuring of State Owned Enterprises (SOEs) has meant that failing enterprises throughout China, including many mines, have had to lay off thousands of employees over the last few years; many other workers are not receiving their wages on time, if at all. The official trade union (All-China Federation of Trade Unions or ACFTU) is government-controlled and has not demonstrated much ability to defend the rights of its workers. Independent trade unions are illegal in China and labour activists continue to face the risk of imprisonment.[42] As far as TIN is aware, there have been no reports of major labour unrest in Tibetan areas relating to mine conditions and pay, although TIN has received reports of wages not being paid on time, or not at all.

The wages of workers can often be influenced by the profitability of the enterprise where they work. A worker at a gold factory in Rebgong talked about the benefits for workers when the factory was doing well. However, he also said that when the production quota was not met the workers did not receive even their basic salaries:

41 According to Article 7 of the International Covenant on Economic, Social and Cultural Rights (ICESCR), which China has ratified: "*The States Parties to the present Covenant recognise the right of everyone to the enjoyment of just and favourable conditions of work which ensure, in particular: (a) Remuneration which provides all workers, as a minimum, with: (i) Fair wages and equal remuneration for work of equal value without distinction of any kind [...]*".

42 See, for example, the Amnesty International report: 'People's Republic of China: Labour unrest and the suppression of the rights to freedom of association and expression', Amnesty International, 30 April 2002 (AI Index: ASA17/015/2002).

> *"They say there is a plan. If the plan is surpassed, the workers receive salary increase and bonuses. During the year [1997] that I worked in the factory the workers were given 800 yuan (US$96) each as a bonus. That year was the best year in terms of production. The production in 1998 was not good. We weren't paid for about four months."*[43]

Workers in China's coal mines have been hit particularly hard by economic reforms and the subsequent problems facing SOEs. In what is perhaps a reflection of this broader situation, one Tibetan, who worked as a driver for a state coal mine in Tsonub M&TAP, reported that in 1997 workers were not paid for four months.

> *"They were extracting coal in the mountain. I had to bring the coal from the mountain to the heaping ground. We [also] transported coal to the county town. The more coal you could transport, the more wages you would receive. I got about 400 yuan (US$48) per month. Generally it was said that the salary was 400 yuan, but later they told people that they would be paid according to the trips they made. If they couldn't work much they wouldn't get a good salary. But later many people were not getting their salary on time. [In 1997] I was not paid my salary for about four months. They said that they couldn't pay and none of us got our salary."*[44]

Truck in mining area **© TIN**

43 TINDoc 539(jw)
44 TINDoc 643(jw)

After four months of receiving no pay he decided to leave the mine to find work elsewhere. He says that he never received the money owed to him from the mine.

> *"Later, when I heard that they were paying that salary I took a few days leave [from my new job] and went to get the salary, but they said my salary wasn't there. When I asked them why, they said that I should go to the personnel office and ask them. When I went to the personnel office they told me that they hadn't received any information about me and couldn't pay me. They said that I hadn't done any work."*[45]

Another complaint that this driver had, which has been echoed by other Tibetans who have worked in salaried jobs in the state mining industry, was that too many costs were deducted from his salary (when it was paid). This does vary from area to area and among different enterprises; this driver said that in his second job, where he was also driving a truck for a mine, he was much better off:

> *"The [basic] salary [in the coal mine] was good, but they deducted a lot of money for meals and rent. They would give us an outfit and then deduct the cost of the outfit from the salary. Therefore in reality we were not getting much money in our hands. In my second job they did not deduct much."*[46]

While living conditions for small-scale workers, particularly artisanal miners who live in makeshift camps, are often very harsh, they are generally better in the larger SOEs. Provision of benefits, such as medical care, housing and food varies. A Tibetan from Chamdo says that when he went to Lhasa in 1994 in search of work he was able to find a job thanks to his uncle who worked in a mining office:

> *"We were about 500 to 600 workers, both Chinese and Tibetan. Workers were mostly young (17-18) and would not stay long as the work is very hard. Most of the Tibetans were farmers. We received free accommodation and food; I was paid 200 yuan (US$24) a month for six hours work a day, six days a week. If one of us was injured during the work, half of the hospital expenses would be paid by the office. There was a small poor hospital at the mine site with both Tibetan and Chinese doctors, but good medicine is very expensive."*[47]

45 Ibid.
46 Ibid.
47 TINDoc R11(RM2)

Accommodation is usually offered to full time workers in larger state mining enterprises – not unusual in China, where mines and factories, functioning as a person's work unit (*danwei*), would usually be a home as much as work place. Although this is changing with the growth in private enterprise and the increasing fluidity of labour migration throughout China, many mining SOEs in Tibetan areas have in effect become small towns. The remoteness of many of these enterprises has required the creation of an urban infrastructure from scratch. For example, the development of mining in the Tsaidam Basin, Tsonub M&TAP, Qinghai, has literally involved the creation of new towns, populated largely by Chinese immigrants.[48] A worker at the Tsakha (Ch: Chaka) salt mine in Wulan county said that accommodation for workers at the factory, which employs few Tibetans, was free:

> *"I had a relative in the salt-mine. He made connections there and I got the job. I had 700 yuan per month for the salary. We don't have to pay room rent. The government gives the rooms. Even for the meals we have a communal mess. We can also cook our own food."*[49]

The large-scale SOEs with many employees, like Chaka salt factory (primarily Chinese workers) and Norbusa chromite mine (primarily Tibetan workers) provide facilities for workers that can include schools, crèches, cinemas and medical clinics or small hospitals. Norbusa chromite mine's publicity material claims that such facilities have been provided for workers:

> *"There is a primary school for children of the workers [and] the mining area and headquarters of the mine have kindergartens, canteens, cinemas, dance halls and baths. The housing of the workers averages 18 square metres per person. With the development of the market economy there have appeared a number of shops, tea houses, restaurants and the commercial establishments run by private individuals. The mining area has assumed the scale of a trading town."*[50]

48 See Chapter 3, pp. 87-88
49 TINDoc 7(rp)
50 'Rhobusa of Tibet...', op cit.

Health and safety

Health and safety conditions in China's mines are notoriously poor. Internationally, the PRC has one of the worst worker safety records for the mining industry (alongside South Africa) with a particularly high death toll in its coalmines, estimated at about 5,000 miners per year in recent years. By the end of the 1990s, many mines, particularly those operating as Township and Village Enterprises (TVEs), were officially ordered to stop production until they passed environmental and health safety standards; many continued to operate regardless.[51] In response to a sustained high death rate in China's mines, the Chinese Minister for Land and Resources, Tian Fengshan, announced at the end of 2001 that the authorities were making fresh attempts to enforce health and safety standards in state-run mines. The authorities have also been showing a greater willingness, in a few cases at least, to hold management of mines and local officials accountable for standards by pursuing criminal prosecutions.[52]

As is the case with environmental protection, many enterprises are either unable to meet the high rates of financial investment needed to maintain high standards in health and safety, for example through better technology, or there is an unwillingness to reduce short-term profits. Local officials responsible for supervising health and safety (and environmental) standards are often the same officials responsible for meeting local economic development targets. They may also be running the mine, or at least seeking to gain some personal profit from the venture. Therefore a conflict of interests often exists between official pressure or personal gain and the implementation of better health and safety standards.

As far as TIN is aware there have been no official Chinese press reports of major accidents in mines in Tibetan areas. However, it is openly acknowledged that accidents and safety problems in mines are frequently covered up and many go unreported. While there is limited available information on safety standards in mines in Tibetan areas, it is realistic to assume that conditions are unlikely to be better than those in the rest of China. A Xinhua report on a national telephone conference held on 20 April 2002 on safety in non-coal mines stated that there haven't been any 'serious accidents' in the TAR during the first three and a half months of 2002.[53] This would suggest that there may have been accidents in previous years.

51 For example, according to a Xinhua report dated 17 November 2001, 11,882 small coal mines had been shut down in the first ten months of 2001 and the death toll had fallen. A week later reports were emerging that during the space of nine days (14 to 22 November) accidents at six township-operated coal mines (five in Shanxi province, dubbed China's 'sea of coal', and one in Shandong province) had killed a total of at least 113 miners (various Xinhua, November 2001). None of the mines were supposed to be operating when the accidents occurred – they had all been closed down awaiting safety inspections.
52 & 53 See next page

The risks facing miners range from poor work conditions, exposure to high levels of heat and noise, exposure to chemicals and other dangerous substances such as mercury and accidents in the work place caused by poor engineering – for example tunnel collapses – and dangerous machinery. There have also been unconfirmed reports of Tibetan workers involved in infrastructure construction for mines being injured or killed as a result of poor safety standards.

A monk from Payul (Ch: Baiyu) in Kardze described the work and conditions at the gold mine in Payul, where he worked for nearly four months. He said that the workers had to dig deep tunnels that sometimes collapsed. While he was working there three or four people died, caught in a tunnel, and one worker in his own group who was buried under collapsing earth also died.[54]

A Tibetan who worked in the aluminium smelter in Rebgong worked in the room where the aluminium is separated from refined bauxite ore in large vats, known as 'pots'. He was responsible for changing the anodes conducting electricity through the smelting 'pots'.[55] Workers in such electrolytic smelting potrooms are exposed to respiratory problems, including a condition known as 'potroom asthma'.[56] This Tibetan describes the dust that workers are breathing in throughout the day, but he didn't report any respiratory problems himself. He had clearly received no education regarding health risks and did not appear to be concerned by health issues. He left the factory for personal, not health-related, reasons:

52 For example, the owner of one of the Shanxi coal mines mentioned in note 50 (see above), held responsible for the blast, was arrested on 26 November (Xinhua, 27 November 2001). A number of administrative officials, including the county Party secretary and the deputy head of the county government, were suspended from duty 'awaiting further punishment' (Xinhua, 26 November 2001).

53 'Eliminate Mining Safety Problems', Xinhua, 21 April 2002, on www.tibetinfor.com

54 TINDoc 4(tm)

55 The aluminium smelting process involves dissolving alumina (refined bauxite ore) in an electrolytic bath of molten cryolite (sodium aluminium fluoride) within a large carbon or graphite lined steel container known as a 'pot'. An electric current is passed through the electrolyte, flowing between a carbon anode (positive), made of petroleum coke and pitch, and a cathode (negative), formed by the thick carbon or graphite lining of the pot. Molten aluminium is deposited at the bottom of the pot and is siphoned off periodically and taken to a holding furnace. It is then often alloyed with other materials, cleaned and then generally cast. Notes adapted from the International Aluminium Institute website, www.world-aluminium.org

56 Winder, C; Jeung, P, 'Health problems in aluminium smelter workers: hazards, exposures and respiratory disease', abstract of article appearing in Journal of Occupational Health and Safety: Australia and New Zealand, No.5, pp.391-402. See Australia's National Occupation Health and Safety Commission website www.nohsc.gov.au

57 (See next page) The aluminium smelting process is continuous. According to the International Aluminium Institute: "*A smelter cannot easily be stopped and restarted. If production is interrupted by a power supply failure of more than four hours, the metal in the pots will solidify, often requiring an expensive rebuilding process.*" Notes on aluminium smelting on www.world-aluminium.org

"The dust [in the smelting room] was like a fog. If you were near to each other you could see each other's faces. If you were far away you couldn't see each other. It was like this inside the factory. The dust was always there. The workers work continuously in shifts. The aluminium is boiled all the time. It never cools down. They say in the factory that if the aluminium is allowed to cool down for one moment, then the cost is about 80,000 yuan (US$9664).[57] *Each day you got to shower. Normally Tibetans don't wash themselves like that, but if you didn't wash your face would be black from the dust. Inside the factory you had to change all clothes apart from your underpants. The factory gives working clothes."*

Although he did not report any negative respiratory effects from his work, he did note that the workers' hair would turn yellow which suggests that some sort of bleaching agent was used in the smelting process at this factory. He also said that the watches of workers would not work if they wore them in the factory – a result of the powerful electric currents in the pot-room:

"All those who go into the factory have their head hair turn yellow. Your hair doesn't turn yellow after having been in the factory for one day. After having worked in the factory for one or two months, your hair turns yellow. We always dyed our hair black. No one from the workforce talked about this [hair turning yellow]. Inside the factory we workers talked about the dust being harmful, but I don't know what was real or not. Apart from the things I really experienced myself, such as your hair turning yellow and your wristwatch getting spoiled, I wasn't aware of other problems."[58]

57 See footnote on previous page
58 TINDoc 746(jw)
59 (See next page) TINDoc 446(jw)
60 (See next page) According to a report on small-scale gold mining mining by the Panos Institute, small-scale miners using mercury to separate the gold from the earth often put their health at serious risk through ignorance of the dangers of mercury. The report explains that liquid mercury is poured over the crushed ore in a pan or sluice to form an amalgam. This is pressed by hand through a cloth to remove the excess mercury. The miners crowd round to light it and the mercury burns off as a white vapour and is inhaled by the miners. The mercury then enters the body through inhalation, ingestion and skin absorption, much of which is passed in urine. However, according to Panos, high exposure can result in the metal being deposited in the central nervous system and the brain. Chronic poisoning can lead to insomnia, severe tremors, brain damage, even death, and by entering the placenta, it can cause birth defects. 'The Lure of Gold – How Golden is the Future?', Panos Media Briefing No.19/May 1996. See the Panos Institute website for the full text of this report www.panos.org.uk

A Tibetan from Lithang who ran a small gold mine said that he and other prospectors who were renting land from the authorities for gold mining had to sign a life insurance agreement with workers before mining started, to agree the level of compensation in the event of any deaths. While he says that it meant that his workers wouldn't create problems if someone died, the local authorities were probably also keen to limit their own liability.

> *"It was made clear beforehand how much money we had to pay in case one of our workers died in the mine. We had to pay 30,000 yuan (US$3614) per person. The workers accepted this regulation and wouldn't create problems if someone died. We signed a document on this to the office responsible for mining."*[59]

Health and safety standards in illegal mining are clearly non-existent and prospectors are often working and living in terrible and dangerous conditions.[60] According to the Workers' Daily, among the thousands of people, mainly Hui Muslims, who flooded to Nagchu (Ch: Naqu) in 1994 following the announcement of the discovery of a large gold vein, many fell ill from diseases such as emphysema.[61] The newspaper said that these prospectors were digging 14 hours a day in harsh weather and dangerous mountain terrain, subsisting mainly on steamed corn. One in 10 of the gold seekers, according to the report, was under 14 years old.[62]

Some of the worst health and safety conditions are those found in China's labour camps. For example, on 18 May 2002, a flood at a remote coal mine in Qinglongzui, near Chengdu, trapped and killed 39 miners, all criminal prisoners. The previous November, 19 prisoners were killed in a similar incident at another of Sichuan's coal mines.[63] In the labour camps and prisons in Tibetan areas, similar to conditions throughout China, abuse and maltreatment is common and forced labour conditions are harsh (see next page).

61 Asahi Shimbun, 8 May 1995. Emphysema is a chronic lung disease.

62 According to the UN Convention on the Rights of the Child, (entered into force 2 September 1990; ratified by China), States Parties are responsible for ensuring that the rights of the child in labour are not violated:

Article 32

1. States Parties recognise the right of the child to be protected from economic exploitation and from performing any work that is likely to be hazardous or to interfere with the child's education, or to be harmful to the child's health or physical, mental, spiritual, moral or social development.

2. States Parties shall take legislative, administrative, social and educational measures to ensure the implementation of the present article. [...]

63 John Chan, 'Prisoners die in Chinese mines', Word Socialist Website, citing press agencies.

Compulsory labour

Compulsory labour continues to be imposed on local communities throughout China as a means of obtaining cheap labour for economic development, as well as constituting the basis of the PRC's judicial and administrative penal systems, including the punishment of political prisoners and labour activists. Over the past 50 years, local communities and prisoners have been an important source of manual labour within mining enterprises in Tibetan areas and for preparing the supporting infrastructure (roads, power, housing). In the Tsaidam Basin in Qinghai province, the most developed mining area in Tibet, the mining industry was built on the labour of forced immigrants and prisoners who also developed the necessary agricultural base to support the new mining towns and settlements that have been established since the 1950s.

China is a member of the International Labour Organisation (ILO), which is generally seen as setting the basic standards in labour rights. However, it remains one of a handful of the ILO's 175 members that have not yet ratified the ILO's two conventions on forced labour: The Forced Labour Convention, 1930 (C29) and the Abolition of Forced Labour Convention, 1957 (C105).

The ILO defines forced labour in the first of these conventions as "*all work or service which is exacted from any person under the menace of any penalty and for which the said person has not offered himself voluntarily*" (C29, Article 2). The main objective of the second convention (C105) appears to have been to delineate more clearly circumstances under which forced or compulsory labour is banned:

C105, Article 1

Each Member of the International Labour Organisation which ratifies this Convention undertakes to suppress and not to make use of any form of forced or compulsory labour

(a) as a means of political coercion or education or as a punishment for holding or expressing political views or views ideologically opposed to the established political, social or economic system;

(b) as a method of mobilising and using labour for purposes of economic development;

(c) as a means of labour discipline;

(d) as a punishment for having participated in strikes;

(e) as a means of racial, social, national or religious discrimination.

According to the ILO, prison labour is only allowed where it is "*carried out under the supervision and control of a public authority and that the said person is not hired to or placed at the disposal of private individuals, companies or associations*" (C29, Article 2c). Article 21 of C29 stipulates that: "*Forced or compulsory labour shall not be used for work underground in mines.*"

Forced/prison labour

China's prison system is based on labour. According to the PRC Prison Law, any prisoner who is able must participate in labour (Article 69). Prisons *(jianyu)* and reform through labour centres (*laogai*) hold prisoners who have been criminally sentenced through the judicial system.[64] A third category of detention facility consists of the re-education through labour camps (*laojiao*), first introduced in 1955 and given legal status in 1957,[65] where people can be sent for up to three years, without judicial procedure, having been administratively sentenced by officials from the Bureau of Re-Education Through Labour. Labour is also carried out by detainees held at PSB detention centres.[66]

Many of the prisoners who have been forced to labour in the PRC's prisons, *laogai* and *laojiao* over the past 50 years are considered by international standards to have been political detainees. During the 1950s and 1960s, tens of thousands of intellectuals, students, religious personnel and others were sent to labour camps and many others were forced to move to remote regions of the country to establish agricultural and industrial settlements. Since the post-Mao era reforms, political prisoners continue to contribute to national construction efforts, including mining, through forced labour. The current Strike Hard Campaign, launched in April 2001, has resulted in a new wave of arrests throughout China. Many of those detained, including large numbers of *Falun Gong* practitioners,[67] have been administratively sentenced and sent to *laojiao* for re-education. A considerable number of current detainees in China, including Uighur Muslims and Chinese labour activists are considered political prisoners.

Historically, the development of the mining industry in Tibetan areas of the PRC was dependent on the labour of criminal and political prisoners in the prison and *laogai* network. Following the suppression of the Tibetan uprising in 1959, many Tibetans, including large numbers of monks, were sent to labour camps in Tibet or elsewhere in China. Although these camps were ostensibly places for ideological re-education they were also a source of cheap labour, largely focusing on road, bridge and dam building, mining and reclamation of land for agriculture. They formed the backbone of industrial development in much of western China, most notably in Qinghai province.

64 China has, in theory, as part of reforming its Criminal Code and Criminal Procedure Law, discontinued the use of the term and institution 'reform through labour'. There are few indications, though, that either the practice or the *laogai* themselves are being phased out. Steven D. Marshall, 'Hostile Elements', TIN, March 1999

65 'Re-education through Labour', China Labour Bulletin, Issue #38, Sep-Oct 1997

66 Detainees are held in PSB detention centres during police, procuratorial and court investigations and either processed and sentenced, or released without process. Many detainees sentenced to administrative detention serve their sentences at PSB detention centres. Steven D. Marshall, 'Hostile Elements', TIN, March 1999

67 Amnesty International had records of nearly 1600 cases of detention, arrest or sentencing of Falun Gong practitioners between June 1999 and March 2000 ('The crackdown on *Falun Gong* and other so-called 'heretical organisations', 'Amnesty International, 23 March 2000). *Falun Gong* supporters claim that up to 10,000 adherents have been sentenced to re-education through labour.

According to the Tibetan historian Tsering Shakya, two of the largest labour camps in the TAR were Nachenbag, where prisoners were forced to work in the construction of a hydro-electricity plant, and the Jangthang borax mine (Tib: Jang Tsala Karpo).[68] A Tibetan monk, who was sent to the borax mine[69] for forced labour after his monastery was closed, talked about the different people working at this mine, including prisoners, political prisoners and monks like himself who were not 'prisoners', but who were engaged in compulsory labour:

> *"In 1959 our monastery was destroyed. The monks were expelled from the monastery, and some monks were educated, some monks were beaten and some monks were imprisoned. And some monks were sent to the countryside of Jang for extracting tsha la [borax]. I went there in 1960, about September. Three years I spent there. We were about 500 monks, and most of these monks were Khampas and Amdowas from about 15 monasteries from the eastern part of Tibet. [We worked] eight hours per day, four hours before lunch and four hours after lunch. [There were also] Chinese prisoners and lay people working there. Prisoners who were imprisoned in 1959 [following the Lhasa Uprising] were also there. Many prisoners died while working there."*[70]

Qinghai province is said to have the highest concentration of *laogai* in China, and its industry was largely built on the labour of prisoners and forced immigrants. These *laogai* were established as 'state farms', 'state mines' or 'state factories', rather than clearly designated prisons. Construction of roads, new towns, mines and production facilities was started during the 1950s in Tsonub M&TAP, where the Tsaidam Basin, rich in minerals, is located. This construction involved thousands of forced immigrants and prisoners – many of them from the east coast of China – working under exceptionally harsh conditions.[71] Forced labour was also used to reclaim wasteland for agriculture to support the developing mining industry. According to Marhsall and Cooke:

68 Tsering Shakya, 'The Dragon in the Land of Snows', London: Pimlico, 1999. The father of Garu nun Ngawang Sangdrol (currently serving the longest sentence of female Tibetan political prisoners), Tashi Namgyal, was among the Tibetans sent to the Jangthang borax mine.

69 Borax: hydrated sodium borate; an ore of boron

70 Although there are no exact figures, hundreds of Tibetans are believed to have died in labour camps.

71 For example, construction of settlements in the mining towns of Mangya and Dahchaidan was started in 1956. See Chapter 3, pp. 87-88

> *"One of the basic purposes of agricultural laogai since their inception in the 1950s was to provide an agricultural support base for official mining activities, a policy applied in Tsonub and also, for example, the Yingguanzhai farms in Mianyang, Kangding [Tib: Dartsedo, Kardze TAP], which supported the Tenpa Tica Mine (Dangdai Ganzi, p.66)."*[72]

Many of these prison camps have now been converted to non-prison use or have been abandoned.[73] However, Columbia University academic and author James Seymour estimates that in 1995 there were still 19 large enterprises (factories and farms) in Qinghai that relied primarily on prisoner labour,[74] a number which accords closely with TIN research.[75] In Xining and Tsoshar, where the highest concentration of *laogai* is to be found, the labour camps today produce a variety of industrial, agricultural and consumer goods. In the remoter, more Tibetan areas,[76] the utilisation of forced labour remains focused more on the production of materials needed for basic infrastructure construction (for example cement), and on natural resource extraction. Prisoners, including political prisoners, work in cement factories, stone quarries, lumber-yards, metal works, and perform dangerous work in resource extraction.[77]

Since TIN started collecting data on political imprisonment in Tibet in 1987, there has only been one confirmed report of a Tibetan political prisoner being forced to labour in a mine. Tsering Dorje, a 30-year old trader from Kardze county was detained in October 1990 for putting up pro-independence posters and sentenced to 12 years. He was first taken to Kardze's prefectural prison at Minyag (Ch: Xinduqiao), and was then transferred to a gold mining *laogai* in Garthar (Ch: Bamei) in Dawu (Ch: Daofu) county. This mine is part of the network of *laogai* focusing on gold mining located in Dartsedo and Dawu counties and referred to by Tibetans as the 'Rang nga khang'. Believing he would not survive his sentence, he managed to escape in February 1993. According to Tsering Dorje, conditions at the mine were extremely harsh. He says that

72 Marhsall and Cooke, 1997, op cit.

73 Susette Cooke, 'A Tradition of Intellectual Dissent', TIN Special Report, 30 July 1999

74 'New Ghosts, Old Ghosts'1998, M. E. Sharpe, p.153

75 Seymour assesses the total number of Tibetans imprisoned in Qinghai, for any reason, at any type of facility (*laogai*, detention centres, drug addict rehabilitation centres and so on), to have been between 1200 and 1500 in 1995. 'New Ghosts, Old Ghosts', p.168

76 Neither Xining municipality nor Tsoshar prefecture hold Tibetan autonomous status, although this area has important ties to Tibetan history and culture, including the monastery of Kumbum and the birthplace of the 14th Dalai Lama, both in Xining. According to 2000 census data, the populations of Xining and Tsoshar are respectively 5.2 per cent and 9 per cent Tibetan (Qinghai Statisitcal Yearbook, 2001). The data takes into account the transfer of two of Tsoshar's counties to Xining in 2000. Five of the six other prefectures in Qinghai hold Tibetan autonomous status, while the sixth, Tsonub, is a Mongolian and Tibetan Autonomous Prefecture.

77 Steven D. Marshall, 'Chinese *laogai*: a hidden role in 'Developing Tibet'', a paper presented to the conference 'Voices from the *Laogai*: 50 years of surviving China's forced labour camps', 17 – 19 September 1999, American University, Washington DC, USA, organized by the *Laogai* Research Foundation.

prisoners were compelled to fulfil daily excavation quotas despite meagre food and little sleep. If they didn't fulfil their quota they would be beaten. He alleges that Chinese prisoners who had no relatives to supplement food committed suicide in preference to being beaten, starved and worked to death, while other prisoners injured or mutilated themselves to avoid excavation work. Attempted escape was met with extended flogging to the naked body, followed by months of isolation in a small, darkened cell, the main door of which was, according to Tsering Dorje, not opened again until the solitary period was completed.[78]

A large prefectural-level prison and labour camp in Minyag (Ch: Xinduqiao), Kardze TAP. Inset: labour camp in Garthar township, Dawu (Ch: Daofu) county, Kardze TAP. © TOTAR, 1997

78 Steven D. Marshall, 'Chinese *laogai*: a hidden role in 'Developing Tibet". This paper also gives further examples of other sorts of labour undertaken by Tibetan political prisoners, including work in cement factories, the forestry industry and cultivation of cash crops.

Community/compulsory labour

The continued use of compulsory labour throughout China is a great irony of the reform period. When the Communists came to power compulsory labour as a form of tax – corvée labour – was one of practices that was to be proudly abolished in order to free the peasants of China and 'serfs' of Tibet from such burdens imposed on them by their former masters.

Since the implementation of the policy of accelerated development in Tibetan areas from the early 1990s onwards, there has been a rapid increase in indirect (and often illegal) forms of local taxation in rural areas of Tibet. This has included a growing use of compulsory labour to fulfil the main economic development priorities set by the Third Forum on Tibet Work held in 1994: irrigation, mining, and the construction of buildings and roads.[79]

The International Labour Organisation (ILO) prohibits the use of compulsory labour for the purposes of mobilising labour for economic development (C105, Article 1.b). However, in China it is often difficult to draw a distinction between compulsory labour and voluntary community labour. The Chinese authorities are aware of their theoretical opposition to conscripted labour and would always refer to community labour as 'voluntary'.

The ILO defines compulsory labour as "*all work or service which is exacted from any person under the menace of any penalty and for which the said person has not offered himself voluntarily*". Given that it is against the law for local governments or citizens to fail obligations under projects directed by the state,[80] local people have little choice but to comply with authorities' requests or instructions to engage in construction work for the general betterment of the state. Local people are obliged to offer themselves 'voluntarily' to take part in national construction efforts; refusal to do so can be viewed as recalcitrant. For example, locals were sent to work on road construction in Gyama township, Maldrogongkar county, Lhasa municipality, for the multi-metals mine, one of the TAR's key mining projects.[81] There are conflicting reports as to whether the locals were paid for their labour. However, there is known to have been opposition to the mine within the local community. Participation in infrastructure construction for the mine is reported to have been obligatory.[82] Under such circumstances it is difficult to judge the extent to which community labour can be considered genuinely voluntary.

79 'Cutting off the Serpent's Head', TIN and HRW/Asia, 1996. The use of compulsory labour is not unique to Tibet. According to Jasper Becker, peasants throughout China can be ordered by local officials to perform a months corvée labour for nothing. Jasper Becker 'The Chinese', John Murray: London, 2000

80 See Chapter 3, p. 80. See also Chapter 6, p. 191-192

81 See Chapter 1, p. 34

82 See Chapter 3, p. 110

Within the local community the distinction can also be blurred. Unpaid work on road construction, for example, is not always resented if it is perceived as being of benefit to the community. Paid community work, for example in a mine, even if obligatory can offer much-needed extra income. This can make it very difficult to assess reports of compulsory labour.

For example, many Tibetans in Kardze TAP are known to be voluntarily involved in small-scale gold mining. However, there have also been unconfirmed reports that Tibetans living in villages near gold-rich areas of the prefecture were obliged by local authorities to engage in prospecting. A monk from Payul, who himself worked in the gold mines in the early 1990s, says that each family had to send someone to work in the mines. According to the monk, the mining land was divided between counties (Payul, Dege and Kardze) and townships. Each family was given a spot in the mining land where they had to send one person to dig for gold for one or two months depending on the size of the family. If they refused, he alleges, they would be fined. The workers were paid dependent on the amount of gold that they found.[83] Many Tibetans do not appear to have been forced to work in the gold mines against their will; the extra income was very welcome. However, if this labour was an obligation, as suggested by this monk and other unconfirmed reports,[84] any family that did not wish to be involved, for whatever reason, would be compelled to participate under the menace of penalty, thereby constituting forced labour conditions.[85]

Conversely, community labour that is paid for, albeit at a low wage, can be resented where limited local benefit is apparent. Where construction is seen as primarily benefiting the state, the local authorities or immigrants over local inhabitants, and there is also no remuneration for local labourers, the resentment can be even higher.[86] A Tibetan farmer who left Tibet in 1994 said that there was a mine about six hours walk away from his village in Dengchen (Ch: Dingqing) county, Chamdo prefecture.

83 The monk says that he and his group were given 800 yuan per person for the gold they found in four months. However, the government had supplied the workers with food for which they subsequently had to pay 300 yuan.

84 See also Marshall and Cooke, 1997, op cit.

85 Mining is central to Payul's economy. Official sources state that gold prospecting has been increasingly developed at rural town enterprise and village organisation levels since 1990. According to the 'Ganzi Annals' (1997), Payul's reserves of rock gold are "*among the most plentiful in the prefecture*". It goes on to say: "*Many nuggets have been found – 12 weighing more than half a kilo have been given to the state, the largest being 6.3 kilos*". (Kardze Annals Editional Committee, 'Kardze Annals', Chengdu: Sichuan People's Publishing House, 1997

86 The opening up of remote Tibetan areas to resource exploitation is a key aim of economic development, including infrastructure construction. Tibetans appear to be aware of these official priorities. A few have reported suspicions that unpaid community labour used in local road construction, ostensibly to benefit locals, is actually serving the interests of the authorities in opening up new mining areas. These kinds of suspicions appear to be aroused when Tibetans have seen what they believe are prospecting teams in the area. TIN is unable to confirm these reports, but they indicate a level of distrust of government motivations.

Tibetan villagers, including himself, were periodically called to work at the mine when their labour was needed. They were paid two yuan (US$0.2) per day and had to take their own tools when they went to the mine. They were also required to take pack animals with them to transport the ore back from the mine down to the road where it was transferred to trucks:

> *"Most of the workers are Chinese but sometimes they call the Tibetan villagers up to work. If they are called they have to go. There are meetings [in the village], mostly about mining. We are told that we have to go and take our animals to the mine. The Chinese have appointed a leader among the Tibetans who speaks from time to time, but the [other] Tibetans only listen – they have no right to speak! The work starts at 7am and we work until midday. After lunch we start again at 2pm and work until 6pm. One time I couldn't go when I was called, so the next time I was called I had to work two weeks instead of one."*[87]

He says that they stayed in tents at the mine site, where work was carried out in teams of three, with two Tibetans digging and a Chinese labourer sorting the rock. Resentment against obligatory labour in the mine appears to have been exacerbated for this farmer by the fact that the mine had also brought many Chinese into the area, who he believes had considerably better living conditions than local Tibetans.

Gold-miners' tents, Kardze TAP **© Anders Højmark Andersen, 1994**

87 TINDoc 1(kr)

Najartse mine near Gyantse (Ch: Jiangzi), Shigatse (Ch: Rigaze) prefecture, TAR © TIN, 1996

5. Environmental issues

The environmental impact of mining on the global eco-system is an issue of growing international urgency and increasing attention is being drawn to the conflict between economic development and environmental protection. Environmental damage caused by mineral resource exploitation is a serious problem throughout China. Beijing maintains that although Tibetan areas have suffered environmental damage, they have generally avoided much of the environmental destruction and pollution faced by most of industrialised China due to their 'backward' state of development. However, the environmental impact that development has so far had on the Tibetan plateau gives some indication of the potential scale of the problem if Tibet is to be developed according to central plans. China continues to pursue a development model in Tibet based on the exploitation of natural resources.

Yet amongst China's leaders there is a growing awareness of the need to act to protect the environment throughout the country and of the specific importance of the Tibetan plateau's fragile eco-system. The headwaters of China's great rivers are found in Tibet and the 1998 Yangtse flood disaster, linked to deforestation in Tibet, has amplified the voices of those concerned with environmental deterioration. The environment is one of the few issues in China today that can be discussed relatively openly. The central authorities have admitted that the mining industry has caused serious environmental problems in Tibetan areas, pointing to the irrational use of resources, pollution and land degradation as key problems.

The first section of this chapter outlines the environmental impact that mining has already had on the Tibetan plateau, looking specifically at land degradation, pollution and the harm mining has caused to livestock and wildlife biodiversity. The second section explores the reasons why, despite the greater attention now being paid to the importance of protecting Tibet's environment and improvements in China's environmental legislation, policy implementation remains limited. It looks at the conflicts of interest that exist between development and environmental protection at both central and local levels, and the fact that environmental protection continues to lose out to what are seen as the weightier priorities of economic development and the political imperative of stability. This section also notes that despite their increasing awareness of green issues, the central authorities remain unwilling to acknowledge the role that central policies continue to play in damaging the environment. The last section provides a case study of the environmental impacts of gold mining in Qinghai province. Gold mining appears to have had the widest reach of all mineral exploitation in Tibetan areas due to the fact that gold is relatively cheap and easy to extract. This case study serves to illustrate the points raised in the preceding sections.

Environmental damage

"The environment of most mines [in China] are [sic] seriously polluted and damaged, which [is] bringing major negative influences to local natural ecological environment, social economy and life of residents".
People's Republic of China Ministry of Land and Resources.[1]

China openly acknowledges the negative impact of mining on its environment. A People's Daily report on China's participation in the 33rd World Earth Day in April 2002 outlined the devastating impact of mining across the country.[2] According to the article, mining activities have caused land collapses at more than 180 sites across China affecting more than 1,150 square km of land. Chinese mining ventures produce 13.38 billion tonnes of solid waste each year, with less than seven per cent of this waste being treated.[3] About 20,000 square km of land have been destroyed or occupied by open-cast mining and the stock-piling of waste residue. The area affected is increasing by 200 square km every year. Waste water released from mines makes up 10 per cent of the national total of industrial waste water and the treatment rate is only four per cent.[4]

Significant damage has already been done to both land and water resources on the Tibetan plateau, including land degradation and pollution. Environmental risks associated with mining are present at the mine sites where minerals are extracted and in associated processing plants.

Most of the information available on the environmental impact of mining in Tibetan areas is concerned with small-scale alluvial gold mining (i.e. the mining of gold found in sediment that has been deposited by flowing water). Chinese press reports have listed the environmental destruction caused by alluvial gold mining in Tibetan and neighbouring areas.[5]

1'MLR Bulletin', March 2000, Ministry of Land and Resources (MLR) website, www.mlr.gov.cn
2 'China marks Earth Day with focus on mining', People's Daily, 22 April 2002
3 Mines produce large volumes of waste. Depending on how this waste is managed and treated it can have a profound effect on surrounding eco-systems, including the potential to become a permanent source of pollutants to natural water systems. ('Breaking New Ground: The Report of the Mining, Minerals and Sustainable Development Project', MMSD, Earthscan: London, May 2002, p.234). Mine wastes are produced in a number of different categories, including overburden (the soil and rock that must be removed to gain access to a mineral resource), waste rock (rock that does not contain enough mineral to be of economic interest), tailings (a residual slurry of ground-up ore that remains after minerals have been largely extracted), and heap leach spent ore (the rock remaining in a heap leach facility after the recovery of the minerals). (Ibid.)
4 People's Daily, 22 April 2002 op cit.
5 See for example, 'Exploring the corridor of Gold', Qinghai Daily, 18 January 1999, 'China's number one gold mine suffers fevered gold rush', Lanzhou Morning News, 10 January 2001, 'Chumarleb completely bans gold digging, fences land and cultivates grass', Qinghai Daily, 25 April 2002, 'Ledu outlaws illegal gold digging to protect the environment', Xihai Metropolitan News, 16 January 2001 and 'Qinghai closes nature reserve', China Daily, 30 December 1999

This includes:

- accidents and damage to land from digging, including destruction of mountain slopes and grass cover;
- great loss and waste of natural resources;
- use of dangerous chemicals (cyanide and mercury) causing serious pollution of the natural environment, including a negative impact on crop growth and the contamination of people and animals' drinking water;
- the disturbance of groundwater systems (which can lead to the drying up of springs);
- natural disasters like land and rockslides and destruction of wildlife habitats.
- Artisanal miners have also been held responsible for the decimation of wildlife including endangered species.

The limited information that does exist supports the assertion that environmental safety standards in these medium to large-scale mines are no better than the generally poor standards throughout China.

Excavations across the Tibetan grasslands for gold mining **© TIN, 2000**

Key Resources

The larger-scale extraction and processing enterprises in Tibetan areas include coal, chromite, asbestos, aluminium, boron, potassium, salt, lead and zinc, copper and gold.

- **Coal:** Coal mines were developed in Tibetan areas during the 1950s and 1960s to support industrial development. Some of Qinghai's coal mines are reportedly being closed down now in favour of hydropower resources as they are generally low grade and inefficient. The TAR's first coal mining enterprise was established in the 1960s in the Tumengela coal mining district of Amdo (Ch: Anduo) county, Nagchu (Ch: Naqu) prefecture. Construction of the 'Machala Coal Mine' in Riwoche (Ch: Leiwuji) county, Chamdo (Ch: Changdu) prefecture, TAR, was listed as one of the priority projects of the TAR 'Specialist Plan' before 2000. It was being developed principally to provide coal to power the Chamdo prefecture cement factory. (TAR National Land Specialist Plan, 1996 – 2020, internal [*neibu*] document, undated. See Chapter 1, note 8. Referred to below as 'TAR Specialist Plan')
- **Chromite:** The Norbusa (Ch: Luobusa) Chromite mine in Norbusa township, Chusum (Ch: Qusong) county, Lhoka (Ch: Shannan) prefecture, TAR, is the largest chromite mine in China.
- **Asbestos:** The Mangya Asbestos Mine in Tsonub (Haixi) M&TAP, Qinghai is one of the largest in China (see note 7, chapter 2).
- **Aluminium:** The Qinghai aluminium smelter in Datong (Tib: Serkhog) Hui and Tu autonomous county, Xining municipality, is one of the largest aluminium producers in China, with an annual output of 100,000 tonnes (www.aluminium.net). According to 2000 census figures, there are 29,160 Tibetans in Datong county, accounting for less than seven per cent of the population; of the remaining population 28.6 per cent are Hui, 10 per cent are Tu and 53.4 per cent are Han. (Qinghai Statistical Yearbook, 2001, p.43). There are also smaller aluminium smelters in Rebgong and Chentsa counties, Malho (Ch: Huangnan) TAP, Qinghai.[6]
- **Boron:** Boron (B) was listed, along with chromium, copper, gold and salt chemicals, as one of the key minerals to be developed in the TAR, according to the TAR Specialist Plan (op cit, p.88). Boron is a non-metallic mineral used in the chemical and building materials industry. The first boron mines in the TAR were developed in the 1950s and 1960s, including the Jangthang Borax (hydrated sodium borate; an ore of boron) mine, one of the largest labour camps in the TAR during this period (see Chapter 4, p. 140).
- **Potassium:** The Cha'erhan salt lake's potassium resources (the largest soluble potassium-magnesium deposit in China) are mined to provide the raw materials for the Qinghai Potash Fertilser Plant (see note 4, Chapter 2). Cha'erhan is located in the Tsaidam Basin, Tsonub M&TAP, Qinghai, which also has significant quantities of other important salt lake minerals including boron, lithium, manganese, sodium and strontium.
- **Salt:** The Tsaidam Basin has 33 salt lakes (People's Daily, 23 November 2001), including Cha'erhan and Chaka. According to the TAR Specialist Plan salt industry processing plants or refineries were to be built in Shigatse (Ch: Rigaze) (where the Zhabuye salt lake is located), Chamdo and Nagchu by 2000 (op cit.).
- **Lead and Zinc:** Lead and zinc mining began at Xitieshan in Tsonub TAP, (between Terlenkha and Golmud) in 1957 and in 1978 a major investment was made under the Sixth Five-Year Plan in upgrading and developing the mines (Marshall and Cooke, p. 1,862). In Tsigorthang (Ch: Xinghai) county, Tsolho TAP an ore dressing plant has been established in the county as a joint venture between Tsigorthang and Zhuhai in Guangdong to process raw materials from the Xinghai Lead and Zinc Mine (Marshall and Cooke, p.2062). Lead and zinc is also mined in Kardze (Ch: Ganzi) TAP, Sichuan and Dechen (Ch: Diqing) TAP, Yunnan (Marshall and Cooke). The Gyama Trikhang multi-metals mine in Maldrogonkar (Ch: Mozhugongka) county, Lhasa municipality is mined for lead and zinc as well as copper.
- **Copper:** See Chapter 1, pp. 28-35
- **Gold:** See Chapter 1, pp. 22-27

It should be noted that another issue of concern, outside the scope of this report, is the impact of associated infrastructure development, particularly power, on the environment. The rich hydropower resources of Tibetan areas are seen as an important energy source for mining and industrial development in both eastern and western China. Several large-scale dams have been built or planned in Tibetan areas to provide power for the mining industry, some of them located on important river courses (see Chapter 2, p. 43). Another serious environmental consequence of current mining practices in Tibetan areas is the irrational use of resources from extensive rather than intensive mining practices. The waste of irreplaceable mineral resources poses a threat to the sustainability of global development. While this issue is mentioned in the context of this chapter, it has already been dealt with in detail earlier in this report. Readers are asked to refer to Chapter 2 (pp. 57-58).

Land degradation

The impact of mining on the land is the most obvious form of environmental damage. For both surface (open-pit/open-cast) mining and underground mining large quantities of soil and rock are excavated, altering the chemical balance of earth and rock and destroying the fragile topsoil. The erosion of the grasslands on the Tibetan plateau, and consequent siltation of water courses, is already a serious problem. Not all soil erosion in Tibet is due to mining; there are several reasons for desertification – deforestation, for example, has caused considerable damage. However, in areas that have been extensively mined, either by mechanised extraction or artisanal and small-scale methods, mining has clearly had a massive impact.

Damage to pastureland is one of the most common complaints of Tibetan nomads and herders who live near mining sites.[7] One Tibetan from Rebgong (Ch: Tongren) county, Malho (Ch: Huangnan) Tibet Autonomous Prefecture (TAP) said that it was impossible for nomads to use land that has been mined.

> *"Nomads can't use [the mined area], the area has been turned upside down, grass doesn't grow there. Wherever they are mining they destroy the grass and make it into sand and stone."*[8]

6 The Rebgong (Ch: Tongren) aluminium factory is one of the China's smaller smelters, producing 5,000 tonnes annually, ('Aluminium Industry www-server', www.aluminium.net). The Chentsa (Ch: Jianza) smelter is reported to be a smaller branch of the Rebgong smelter. Huangnan (Tib: Malho) Aluminium Industry Ltd. Co. produces a total of 8,000 tonnes of electrolytic aluminium p/a at 95 per cent and above aluminium content. Under the list of construction projects on the prefecture's official website (www.huangnan.gov.cn) are several upstream and downstream factories related to the aluminium industry.

7 For more details on the concerns and complaints of Tibetans, including attempts to halt mining or seek compensation, see Chapter 3.

8 TINDoc 554(jw)

Some of the most damaging long-term environmental impacts on the Tibetan grasslands from mining are a result of enterprises failing to clean up sites when mining is finished. Forward planning for mine closure, which includes investment in proper restoration of mining sites, is increasingly viewed (from within the international mining industry as well as by environmentalists) to be as important as the exploration and extraction stages.

In an article published in Tibet Daily in December 1996, the general manager of the Nagchu Prefecture Mineral Development Company criticised the short-sighted view of most mineral extractors, stating that their mining methods are "*damaging and predatory*". He goes on:

> *"After mining (for gold), [land] is not refilled or returned to cultivation. This not only causes serious damage and great losses to existing deposits, it also damages geology, landforms, and grasslands, leaving behind the hidden danger of desertification of the grasslands, which has a definite effect on pastoral production."*[9]

Even where the land is backfilled (i.e. the earth that has been dug up is pushed back into the mine pits), soil erosion can still be a problem. The productive topsoil on the Tibetan plateau is thin and once the soil has been excavated and heaped it becomes sterile. In order to maximise the chances of repairing the environment the pits need to be excavated with care to preserve topsoil and then refilled with equal caution to replace topsoil. Further care also needs to be taken to re-vegetate the soil with suitable plants. Even when sites are reported to have been restored after mining there appears to be little evidence that this kind of care has been expected or taken.

Environmental initiatives, like other policies in China, are usually dictated from above. This results in problems that have been common throughout the history of the People's Republic of China (PRC). Firstly, a failure to incorporate local knowledge into policy directives has frequently led to misguided initiatives that are not suited to local conditions. Secondly, the primary aim is to meet targets and quotas set by higher levels, rather than to address actual problems. For example, Chinese press reports on the backfilling of land in Chumarleb (Ch: Qumalai) county, Qinghai, following widespread gold prospecting, give no indication that attention has been paid to how this should be done. It appears that pits have simply been filled in and grass replanted on the orders of higher authorities.[10] The degree to which these attempts to re-vegetate the area are successful in the long-term will give some indication of the level of care that has been taken to repair the damage.

9 'The search for countermeasures to develop mining in Northern Tibet [*Zangbei*]', Tibet Daily, 4 December 1996
10 See pp. 171-179 for details of gold mining in Chumarleb.

Alluvial gold mining is carried out either by panning for gold in river-beds or digging for deposits that have collected under the surface of pastureland. In both cases considerable damage is done to the environment in the process. Artisanal and small-scale mining often affects a much greater land area than the sections of pastureland being mined; the surrounding area can be left marked by thousands of small holes, indiscriminately dug out in search of minerals, most commonly gold, or for camp use.

Although there are many species of grass in the inter-linking micro-ecosystems on the Tibetan plateau, most have very few roots and their continued growth is naturally fragile. In areas where seasonal permafrost freezes and thaws the soil, this can start a die-back of the grasses surrounding holes dug out for exploration or mining, and the holes continue to enlarge as rivulets form. One effect of this is to create ideal habitat for rodents who burrow under the grasslands causing further damage. The lack of predators due to the human presence in mining areas also means that the natural cull of rodent populations is disturbed and they can multiply abnormally. The result can be extensive and spreading damage.

A Tibetan from Lithang (Ch: Litang) county, Kardze (Ch: Ganzi) TAP, Sichuan, who ran a small-scale private gold mine described subsidence of land where he had mined:

> *"We make small holes; we dig directly down into the rock. Then when we have reached inside the rock we excavate sideways in order to take out the gold. The [tunnel] is narrow at the mouth and large inside below the ground. When there is a lot of rain in summer, the wooden construction inside the hole rots down and then the pastures are damaged a lot because the ground sinks in".*[11]

Steven Marshall and Susette Cooke, who carried out extensive fieldwork in Tibetan areas outside the TAR in the mid-1990s, described the 'devastating' effect of small-scale gold mining along the rivers in Kardze TAP. In a stretch several kilometres long from Garthar (Ch: Bamei) township in Dawu (Ch: Daofu) county towards the prefectural capital Dartsedo (Ch: Kangding), "*the river's bed and banks have been intensively mined, leaving behind a ravaged channel of deep pits and mounds of rock and gravel.*"[12] In Dege county, where they saw a heavy concentration of mining camps Marshall and Cooke reported that "*none of the many abandoned camps had any repairs done to damaged river beds and banks*". They commented: "*Deep pits, high bars of gravel and disrupted river flow is not only environmentally destructive, but a threat to people and animals living in the vicinity*".[13]

11 TINDoc 446(jw)
12 Steven D. Marshall and Susette Cooke, 'Tibet Outside the TAR', 1997
13 Ibid.

Pollution

Air and water pollution are both serious environmental problems associated with the mining industry, particularly when regulation of environmental protection standards is poor and technology is outdated. Air pollution can include dust from blasting operations; the release of gases during excavation; and the emission of heavy metals, sulphur dioxide and other pollutants, such as fluoride, from smelting operations. Ground and surface water in mining areas are at high risk of contamination from leaks and spills from toxic chemical solutions used during smelting and leaching processes. Waste from mines needs to be treated and capped with impermeable material, or it will react with air and water causing chemical reactions that result in 'acid mine drainage' – the separation of heavy metals[14] from the rock or waste which then contaminate ground water and surface water systems.

Pollution control is virtually non-existent in small-scale and artisanal mines. In 1996, the US Embassy reported a conversation with a foreign mining company official who said that the inadequacy of environmental protection at gold mining sites in western and southern China was 'harrowing':

> *"At many mining sites, there is mercury and cyanide scattered over the mountains.[15] After gold recovery, chemical-laden debris is pushed over the hills into creeks, huge areas are devastated, and villages downstream have no idea of how contaminated the water they use is. In many instances, arsenic accompanies gold, so the arsenic gets scattered around the countryside, along with the cyanide and the mercury."*[16]

Larger mines are also reported to have very low standards, many of them continuing to operate on outdated technology. Chinese scientists Song Xinyu and Yao Jianhua, from the Commission for the Integrated Survey of National Resources, reported on the poor pollution control in the Tsaidam Basin's mines at an International Symposium in Qinghai in 1998. They stated that:

> *"There are few measures to prevent pollution, with the result that wastes pour into the rivers, endangering livestock, contaminating lakes downstream."*[17]

14 Heavy Metals: Metallic elements, including those required for plant and animal nutrition, in trace concentration but which become toxic at higher concentrations. Examples are mercury, chromium, cadmium, and lead. (Glossary of the Energy Information Administration, US Department of Energy on www.eia.doe.gov). Heavy metals are also sometimes referred to as toxic metals.

15 Mercury and cyanide are both chemicals used to separate gold from earth/rock.

16 'Gold Mining in China: Taming the Wild West', US Embassy, China, May 1996

17 Song Xinyu and Yao Jianhua, 'Resources Exploitation and Conservation of Qaidam Basin under the Principle of Sustainable Development', a paper presented to the International Symposium on the Qinghai Tibet-Plateau, Xining, 24 July 1998

Makeshift gold-miners' tent on river-bed, north of Dege county town, Kardze (Ch: Ganzi) TAP © TOTAR, 1997

The processing plant at the gold mine in Nyima township, Machu (Ch: Maqu) county, Kanlho (Ch: Gannan) TAP, Gansu province, uses the cynanide/charcoal process to separate the gold ore.[18] This means that the gold ore is leached from the rock by open-air spraying with cyanide solution. The crushed ore is placed on pads lined with clay or plastic. The cyanide solution dissolves the gold and seeps through to a collection pond, bearing the gold. At the processing plant, the gold is separated from the cyanide solution by being pumped through tanks containing charcoal (charcoal absorbs the gold). The cyanide solution is then transferred to holding ponds and reconstituted and reused.[19] Cyanide is extremely poisonous for people, plants and wildlife and once ingested, inhaled or absorbed even in very small doses is fatal.[20]

While no human death or illness has been reported at the Machu mine, animals have died after drinking contaminated water. The clearest explanation has been that animals have in the past drunk from 'poisonous water' pits, probably a reference to the holding ponds. The mine is reported to have paid nomads 1,000 yuan (US$120) in compensation for each animal that died. A Tibetan nomad from the area said that things had improved recently, with fewer animals dying from drinking contaminated water. He thinks that this is due to a combination of better safety precautions at the mine and greater caution from the herdsmen who have learnt from experience that the water in these pits is toxic:

18 'Tsewang [Ch: Caiwang]: the spirit of Gerke', China Today [China Hoy] Spanish Edition, Volume 38, No.7 July 1997 p.61-63
19 See for example 'The Use of Cyanide in Mining', Mineral Resources Forum (established as an intiative of UNCTAD) www.mineralresourcesforum.org or Dennis T. Trexler, Thomas Flynn and James L. Hendrix, 'Heap Leaching: Introduction', summary of two articles (Trexler 1987 and 1990 published in Geothermal Resources Council Transactions, Vols 11 and 14), published on Geo-heat centre, Oregon Institute of Technology, website www.geoheat.oit.edu
20 'The Lure of Gold – How Golden is the Future?', Panos Media Briefing No.19/May 1996. See the Panos Institute website for the full text of this report www.panos.org.uk

> *"Previously they didn't fence the pits where they stored the poisonous water and there was only one person to guard the area. The herdsmen didn't know that this water was poisonous so they probably didn't see any problem with their animals grazing in that area. Now they have fenced off the water and there are some people that guard the place. Besides, the herdsmen are now taking great care not to let the animals graze around these places."*[21]

However, the same Tibetan also claims that water from the pits seeps into the ground and sometimes overflows into a stream running through the area, again affecting the health of livestock. Other Tibetans from the area say that the quality of the grass has been adversely affected, which could be a consequence of leakage from the contaminated pools if these are not sealed properly or if they periodically overflow. One nomad believes that the gold mining has caused a medicinal spring to dry up. Excavation of earth for mineral exploitation can change underground water patterns leading to the drying up of wells:[22]

> *"Many people in our area recovered from different diseases after drinking this water. Moreover the water was very effective for wounds. This medicinal water was found in a small well but now it has dried up."*[23]

Villagers in Zhethongmon (Ch: Xietongmen) county in Shigatse (Ch: Rigaze) prefecture, are reported to have complained to the prefectural and regional level government about a gold mine polluting the stream they used for drinking water. As a result, the mine management paid for a well, which was built by the villagers, and for a pump.[24]

It is interesting to note that in both of the above cases the respective mines' management took action in response to the complaints of contamination from locals, perhaps due to concerns about attracting further attention from higher authorities. In the first case, in addition to paying compensation to nomads who lost livestock, the mine tightened its safety measures to prevent animals from drinking contaminated water. In the second case, while the mine provided locals with an alternative water source by financing a well and pump, they are not reported to have attempted to deal with the actual environmental problem (contamination of the stream). In this latter case at least, the aim appears to have been to address the villagers' immediate needs rather than to tackle environmental pollution and its long-term consequences.

21 TINDoc 1011(jw)

22 See for example Daniel Giammar, 'Surface coal mining and environmental degradation in the US', June 1997, www.gps.caltech.edu. Although this paper is primarily concerned with coal mining, the general effects of surface and underground mining described are pertinent to all such mining. A report in Lanzhou Morning News (10 January 2001) on mass illegal gold mining in Wen county, Gansu, stated that "*owing to geological structures, springs in the area dried up and people from the upper village groups had to go to a place one kilometre away and load water onto their draft animals.*"

23 TINDoc 1017(jw)

24 TINDoc 15(rg)

Livestock

Among the common complaints from nomads who are living in or near to mining areas have been the injuries and deaths of animals in their herds from blasting, pollution and accidents (animals falling into the holes pockmarking the pastureland). Tibetan nomads also lose livestock to artisanal gold miners, who set up camp on the Tibetan grasslands and poach from the nomadic herds.

A Tibetan from Rebgong in Qinghai talked about the injuries to animals from blasting at a gold mine:

> *"Before, many animals were wounded because they were hit by flying stones caused by the explosions. Later people were afraid that animals would get hurt and didn't send them to that place any more."*[25]

He also said that animals died from contamination of the river from the gold factory:

> *"Sometimes if the animals die, the workers hide the carcasses. If the nomads find out that an animal died [from contamination] they cause problems for the factory. Therefore the workers secretly bury the dead animals to avoid troubles."*[26]

According to the Tibetan from Lithang who ran his own gold mine, many animals in the area died because the pits weren't refilled:

> *"Many animals die. After the work was finished, when it rained, the holes got filled up with water. Then the animals walked into the holes and drowned. This area has become like a place which has been left destroyed by the Chinese army."*[27]

Several villages in the two townships of Mepa and Thogpa in Rebgong county, are reported to have been affected by fluoride pollution from the Rebgong Aluminium Smelter located in Mepa.[28] Reports from the area state that smoke from the factory is causing the teeth of livestock to fall out and that it affects the quality of the grassland and crops. A western ecologist who visited the area said that the smelter is 'extremely unpopular', even amongst those Tibetans living on the surrounding hills where the effect is minimal.

25 TINDoc 554(jw)
26 Ibid.
27 TINDoc 446(jw)
28 For details of the Rebgong smelter see note 6

No commercially feasible process of smelting aluminium has been discovered that does not use fluoride.[29] As a result emissions from aluminium smelters, particularly those using outdated technology, can contain dangerous levels of fluoride. The US Environmental Protection Agency has stated that while fluoride emissions from aluminium smelters do not have significant effects on human health they do have an adverse effect on livestock and vegetation.[30] In certain circumstances, according to the Australian aluminium company Tomago, fluorides in the air can damage vegetation, with some types of plants being more susceptible than others. The company also gives an account of the damage that can be done to livestock:

> *"Cattle grazing on grass subjected to fluoride emissions can experience damage to their teeth, especially when they are young, and can accumulate fluoride in their bones. Cattle are generally regarded as the animal species most sensitive to fluoride."*[31]

Herding livestock on the Tibetan grasslands **© TIN, 2000**

29 'Facts about Fluoride' on Australian company Tomago Aluminium's website: www.tomago.com.au. The introduction of fluoride was a key factor in the success of the original Hall-Heroult method, discovered in 1886, of extracting aluminium metal from alumina (ibid). This method remains the basis for all modern primary aluminium smelting ('Aluminium Smelting', International Aluminium Institute website www.world-aluminium.org). The aluminium smelting process involves dissolving alumina (refined bauxite ore) in an electrolytic bath of molten cryolite (sodium aluminium fluoride). See note 55 Chapter 4

30 'Health and Welfare Effects of Fluoride' in 'Primary Aluminium Draft Guidelines for Control of Emissions from Existing Primary Aluminium Plants', a publication that referred mainly to older plants not equipped with advanced fluoride suppression systems. Cited in 'Facts on Fluoride', Tomago, op cit.

31 'Facts about Fluoride', Tomago, op cit. Like many other minerals, fluoride at low levels can be beneficial. For example, according to Tomago, in areas where fluorinated water is supplied, children's teeth have had fewer cavities and the fluoridation of drinking water is practised by most Australian cities. Fluoride is also used as medication to treat bone deficiency diseases like osteoporosis (ibid).

One Tibetan from the area talked about the impact that pollution from the factory is having on the villagers' crops and livestock. He also expressed an understandable, if groundless, fear that humans will be affected[32] and will also start to lose their teeth:

> *"The biggest problem for the people in our village is the aluminium factory. The smoke from this factory harms our crops [and] because our animals lose their teeth they can't eat grass. Earlier each household used to have 15 or 20 goats while these days there aren't any goats. We earlier had a few goats. Later we sold them thinking they might die. People worry that gradually it will also affect the people and that the people will also lose their teeth. The smoke doesn't go into the air. It stays low and comes to the village. After they make the aluminium it is taken to Lanzhou.[33] In Lanzhou they make utensils and electricity wire out of the aluminium. They take everything to Lanzhou. Our place gets nothing but the smoke."[34]*

A Tibetan who grew up in the area and returned there for a visit at the end of the 1990s said that one of the most striking changes he noticed was the decrease in livestock, particularly sheep and goats. The aluminium factory was widely discussed among the families he visited while he was staying in the area and he understood that many people had sold their animals when some started to lose teeth:

> *"Everyone is talking about the animals they lost and how they are not able to keep animals. The villages near the factory are most affected by its smoke because they are situated in the valley and at the side of the factory between the mountains in such a way that the wind also blows the smoke in their direction. The grass of the pasture turns yellow as a result and so the animals can't eat it. The fields are also harmed by the smoke from the factory and the crops don't grow well."[35]*

He also says that due to a reduction in livestock there is less manure to fertilise the fields and this further damages the crops. Damage to livestock and crops can be devastating for Tibetans, most of whom rely on farming and herding as their main sources of livelihood, with few alternative opportunities available.

32 The Xihai Metropolitan News recently carried a report on a project to reduce excess fluoride in the drinking water of 11 counties in Qinghai: Huangzhong and Datong (Xining municipality), Huzhu, Hualong, Minhe, Ledu and Ping'an (Haidong prefecture), Chentsa and Rebgong (Malho TAP), Mangra (Tsolho TAP) and Nangchen (Yushu TAP). The report stated that: "*fluoride poisoning, which occurs when fluoride in drinking water accumulates in the human body, happens mainly in the eastern parts of Qinghai*" ('Regions with localised fluoride poisoning start water reform projects', Xihai Metropolitan News, 18 July 2002). No link was made in the report between the excess fluoride in drinking water in Chentsa, Rebgong and Datong counties, and the aluminium industry present in these counties (see p. 150).

33 The capital city of Gansu province.

34 TINDoc 616(jw)

35 TINDoc 381(jw)

Wildlife biodiversity

Mining in Tibetan areas has had a negative impact on wildlife biodiversity through the destruction or alteration of habitats and poaching. Many of the prospectors who have entered Tibetan areas in search of gold also hunt. For poor migrant workers, the remoter areas of the Tibetan plateau, with their inhospitable climate and high altitude, hold two main attractions – gold and wildlife. These areas, protected in the past by their remoteness, have become increasingly accessible as a result of road construction and the availability of suitable vehicles. According to naturalist George Schaller commercial hunting, carried out by local nomads, migrant workers, and more recently organised gangs of motorised hunters, has been the main cause for the decline of wildlife on the Tibetan steppe.[36]

The areas of northern TAR and Qinghai, where tens of thousands of prospectors have mined for gold, are also areas of significant importance for Tibet's biodiversity. Some of them fall under official conservation areas, like the Hoh Xil[37] (Ch: Kekexili) in Qinghai and the Jangthang nature reserve in Nagchu and Ngari (Ch: Ali) prefectures, TAR. Miners have been partly responsible through their illegal hunting activities for the serious decline in wildlife in these areas, such as the wild yak, chiru (Tibetan antelope), gazelle, kiang (Tibetan wild ass), blue sheep, argali (sheep), wolf, fox, bear and snow leopard.[38]

According to the Mineral Resources Law, approval from relevant departments is required to mine in nature reserves.[39] This offers little protection against state exploitation of these areas, while there is virtually no control over most of the small-scale exploitation – this has been outside of the law, or quasi-legal at best. While larger state mining enterprises tend to be self-contained, with the needs of workers being met within the confines of the enterprise, artisanal miners and migrant labourers make use of local resources and are more likely to hunt local wildlife.

36 George B. Schaller, 'Wildlife of the Tibetan Steppe', Chicago and London: University of Chicago Press, 1998

37 Mongolian name, meaning 'blue glass'. This is the common transliteration used for the nature reserve. 'Hoh', meaning blue, is transliterated differently elsewhere, for example in 'Kokonor' or 'blue lake', the common transliteration of the Mongolian name for Qinghai Lake.

38 Tibetan nomads are also responsible for the decline in wildlife through hunting. While nomads traditionally hunted wildlife for subsistence, hunting has become increasingly commercial since the 1960s, and Tibetan nomads, along with migrant workers, have become involved in the endangered species trade. Other factors in the decline of wildlife include the recent influx of nomads into areas that were previously uninhabited, for example the Jangthang nature reserve, and the intensification of livestock rearing, for example around the source of the Yellow River in Yushu. See Schaller, 1998, op cit.

39 Mineral Resources Law (1996 amended version; came into force 1 January 1997)

Article 20 Unless approved by the relevant departments in charge authorised by the State Council, no-one may mine mineral resources in the following places:

5) natural reserves and important scenic spots designated by the state, major sites of immovable historical relics and places of historical interest and scenic beauty that are under state protection.

A Qinghai Daily article, describing the environmental damage done by gold prospectors in Chumarleb county Yushu TAP, refers to the decimation of wildlife, including several endangered species:

> *"The gold peasants [jin nongmin] stole and pillaged the animals and property of the pastoral herdsmen and they wantonly poached wild animals so that the Tibetan antelope, the Mongolian gazelle, the white-lipped deer, the wild yak, the wild donkey and the musk ox, which lived in this area in great numbers in former times, have now virtually disappeared."*[40]

George Schaller visited the Hoh Xil nature reserve and the nearby Tuotuohe area in summer 1997, in order to ascertain the status of wildlife. He had heard reports that nomads, gold prospectors and motorised armed gangs of Tibetans and Chinese poachers from as far away as Golmud and Chamdo had decimated the wildlife, especially the chiru (Tibetan antelope) because of its valuable wool, used to make shahtoosh shawls.[41] He reported that the scarcity of wildlife was 'disconcerting'. He described one 75km walk from the Hoh Xil reserve southeast to Tuotuohe:

> *"One aspect of our walk was dispiriting: we looked in vain for a vista with wildlife. Wild yaks are gone, their bountiful meat always in demand, and argalis persist at most as a tiny remnant. We observed a total of 9 gazelles, 11 chiru males, and a lone kiang."*[42]

Tibetan landscape, October 2002 **© TIN**

40 'Exploring the 'Corridor of Gold", Qinghai Daily, 18 January 1999
41 The chiru are an endangered species. Although the international trade in chiru wool is illegal, shahtoosh shawls have been a style symbol in West since 1979, where shawls are sold for $2000 to $8500 each (Schaller, 1998, op cit.)
42 Scaller, 1998, op cit.

Regulation and Control

Over the last two and half decades China is widely acknowledged to have made major advances in developing a legislative framework to regulate environmental protection.[43] Although economic development and social/political stability remain the two key priorities, increasing attention is being paid to the environment in official policy statements and plans. Foreign, international and domestic aid and investment are funding more than 200 environmental and energy projects throughout China.[44] Environmental issues have received increasing attention in the Chinese media, which has been allowed a degree of freedom in its environmental reporting far beyond that permitted for most issues.

However, despite the positive signs on paper, the authorities openly admit that there is still a yawning gap between official regulations and their implementation; and that far from cleaning up the environment, the situation is getting worse, particularly in China's western regions. In December 2001 Xinhua reported on a survey carried out by the State Environmental Protection Administration (SEPA) and the Chinese Academy of Sciences (CAS), according to which the environment of the western regions of China continues to deteriorate:

> *"A survey of environmental situations in 12 provinces in the west found [that] almost none of the old environment worries, such as soil erosion, desertification, and shrinking of forests and grassland, has been successfully curbed. In fact, many of these problems have worsened."*[45]

This difference between policy and implementation is due to various factors, including conflicts of interest at both central and local levels, and low levels of environmental awareness combined with an unwillingness to take responsibility for existing damage and future protection.

43 See for example Yohei Harashima, 'Environmental Governance in Selected Asian Developing Countries', International Review for Environmental Strategies, Vol.1, No.1, pp.193-207, Institute for Global Environmental Strategies, 2000. In this paper Yohei Harashima outlines the key legislative developments starting with the First China National Conference on Environmental Protection, Beijing, 1973 and the enactment of China's Environmental Protection Law (Trial Version) in 1979. By 2000, China had six environmental protection laws, nine laws for resource protection, 28 environmental administrative regulations, 70 rules and 375 national environmental standards, in addition to more than 900 local environmental regulations. The revised Criminal Law (1997) made it a criminal act to destroy the environment and/or natural resources. (ibid.)

44 An 'Inventory of Environmental Work in China' published in China Environment Series Issue 4 (2001), lists 256 projects. The inventory only lists US governmental activities and does not include, for example, British or EU projects. It gives a breakdown of documented projects as follows: US government activities (16 agencies and 80 projects); US and international non-governmental and academic (36 organisations, 76 projects); Chinese and Hong Kong non-governmental activities (12 organisations, 50 projects); multilateral organisation activities (4 organisations and 50 projects). China Environment Series Issue 4 (2001), published by the Environmental Change and Security Project of the Woodrow Wilson Center on www.ecsp.si.edu

45 'Study shows environment in western region worsening', Xinhua, 29 December 2001

Environmental Impact Assessments

Environmentalists and international agencies like the World Bank now consider Environmental Impact Assessments (EIAs) to be essential in the development of any mining enterprise. A draft law on environmental impact assessment was debated at the 19th session of China's Ninth National People's Congress (NPC) Standing Committee on 27 December 2000. According to the People's Daily (28 December 2000), the draft law stipulated not only that EIAs should be carried out for all construction projects, but that governments at and above the city-level should also organise EIAs and submit reports on them before making industrial development and natural resource exploitation policies or plans that might bring about negative environmental consequences.[46]

Although it is not clear whether this law has come into effect yet,[47] the principle of carrying out EIAs for mining projects has already been written into general policy guidelines. China's minister for Land and Resources, Tian Fengshan, said in November 2001 that EIAs and geological disaster assessments must be undertaken before new mines can be opened, as part of a drive to increase management and control over mineral resources.[48] The TAR's Tenth Five-Year Plan (2001-2005) also states that EIAs must be carried out for all projects involving exploitation of natural resources. According to the TAR Statistical Bureau's communiqué on Tibet's 2001 development, environmental impact assessments were carried out for all construction projects (large, medium and small-sized) that commenced during 2001.[49]

46 'Legislators to Legalise Environmental Assessment', People's Daily, 28 December 2000
47 According to Zhang Qingfeng (Chinese State Environmental Protection Administration and World Bank) and Robert Crook (World Bank): 'Chinese policy makers began to discuss environmental management regulations for construction projects that ultimately will lead to the creation of a stronger Environmental Impact Assessment Law in 2002'. 'Environmental Protection in China's Tenth Five-Year Plan', summary of papers given by Zhang and Crook at a US-China Business Council meeting 14 April 2000, published in China Environment Series Issue 4 (2001), op cit.
48 'Illegal mining gets shafted', China Daily, 27 November 2001
49 'Statistical Communiqué on Economic and Social Development of Tibet Autonomous Region in 2001', released by the TAR Statistical Bureau 25 March 2002, published (in Chinese) on the Tibet Daily website 26 March 2002. Translated by BBC Monitoring. The report also stated that 15 'key industrial enterprises' had met pollution control standards within the specified time frame and that three cement production lines had been shut down.

Political will

The political will of the central authorities in Beijing is fundamental to successful implementation of environmental protection regulations. At a central level in China there is a basic conflict of interests between the desire to exploit Tibet's resources and the need to protect the fragile environment of the Tibetan plateau. Awareness of the importance of Tibet's environment increased following the major Yangtse River floods in 1998, blamed in part on deforestation around the upper reaches of the river. The authorities have now accepted that environmental damage on the Tibetan plateau is not simply a problem for local inhabitants; it constitutes a serious threat to the whole country and to other Asian nations.

However, as outlined in Chapter 1, central development policy for Tibetan and other western regions is based on resource exploitation, focusing on the need to meet domestic demand for resources, as well as on the wider political aims of integration and assimilation to ensure stability. The policies of the Western Development Campaign indicate that this basic development policy is unlikely to change in the near future, even though many observers, including economists and environmentalists within China, are critical of an approach to Tibet's development based on resource extraction, believing that 'green' development is more appropriate.[50]

The decision taken among China's top leaders to put their full and personal backing behind a project or policy can be critical to implementation. Prestige projects, usually large-scale engineering feats, like the Three Gorges dam on the Yangtse River, the Golmud-Lhasa railway, and plans to divert water from the Yangste to the Yellow River, are often pushed forward despite convincing environmental and economic arguments against them. An issue as complex and long-term as environmental protection is unlikely to have as much impetus as, for example, the construction of a new dam or the development of a major new resource. Environmental protection is, in theory, an integral part of Western Development. However, it appears that environmental issues remain of secondary concern when pitted against successful completion of the main projects of the campaign – most of which are concerned with large-scale infrastructure construction and resource exploitation. The authorities claim that adequate measures will be taken to protect the environment during construction and operation of these projects, an ambitious promise given China's poor record in environmental safety implementation. But, according to the official line, there is never any question that the projects themselves might not be advisable or even viable on environmental grounds.

62 See Chapter 6 for a discussion on the model of development being imposed in Tibetan areas.

The restrictions still placed on environmental reporting in China illustrate this point. The growing Chinese 'green media' and NGO network have been allowed a level of investigative reporting unheard of in other sectors. Explanations of this have included the fact that investigative reporting of local problems provides support to central attempts to assert control over local officials. However, criticism of Chinese central policy and aims generally remains out of bounds for the media.[51] Chinese environmental journalist Wen Bo pointed out (in a US publication) that despite improvements in environmental reporting the media have failed to inform the public about the impact of dam construction on the environment. He suggests that this is due to the powerful central backing and associated prestige of the Three Gorges dam:

> *"Because China's problems with floods have historically been so severe, every hydroelectric project is hailed as a means to conquer nature and halt flooding. [...] After the Three Gorges Project was approved by Chinese authorities, articles opposed to hydroelectric projects, or their negative environmental impacts, have seldom been allowed to be published in China."*[52]

However, the state is becoming more aware of the long-term economic costs of poor environmental practice. Economic interests can add weight to central concern over environmental problems. As discussed in Chapter 2, efforts to clamp down on mining and reassert central control have developed as the authorities have become increasingly aware of the economic implications of current mining practices, most notably from illegal mining of state resources and the waste of irreplaceable minerals through irrational exploitation.[53]

Economic development vs environmental protection

The various conflicts of interest that exist at a central level – political, economic and environmental – filter down to the lower levels responsible for implementing environmental regulations. The priority accorded to economic over environmental interests is considered to be one of the main reasons for the failure of regional and local officials to implement environmental regulations. Due to the fact that there is insufficient political will at a central level to give real priority to environmental protection on the Tibetan plateau in the face of competing interests, there is limited incentive at a local level to prioritise environmental protection.

51 Academics appear to have been given more leeway in criticising broader Chinese economic policy. For example, see Chapter 6 for critiques by Chinese academics of the economic model being used for the development of Tibetan areas. This is probably due to the limited audience of academic work. The wider public access to media would make such criticism much more sensitive.
52 Wen Bo, 'Greening the Chinese media', China Environment Series, Issue 1, published by the Environmental Change and Security Project of the Woodrow Wilson Centre on www.ecsp.si.edu, fall 1997
53 See Chapter 2. Also see pp. 171-179

Local officials are effectively being ordered to follow contradictory and opposing policies. Tibetan areas have been charged with accelerating economic development and facilitating resource exploitation in line with central demands. Since the 1980s, but particularly during the 1990s, there has been increased pressure on local governments to be entrepreneurial and launch enterprises. The central and regional/provincial governments have encouraged Tibetan areas to develop local government, collective[54] and private mining enterprises as a way of accelerating economic development. One of the consequences of this has been the progression of the attitude of 'development at all costs', exacerbated by the personal interests that officials have in the success of mining enterprises. The subordination of environmental interests within this context has led to serious long-term damage. However, there is now increasing pressure on the same authorities responsible for accelerating economic development to be more circumspect in their use of resources and to spend more money on environmental protection.

In some Tibetan areas mining has become an important source of income for local governments. Where this is the case, the primary concern for local officials is to protect the enterprise and to ensure that it continues to generate as much profit as possible. A large portion of local government spending is on administration, with the wages of government employees being one of the highest costs. Therefore, local mining enterprises can be a principal source of income for paying government staff salaries.

For example, the aluminium factory in Rebgong, although resented by many local Tibetans for polluting the local environment and causing damage to livestock and crops (see pp. 157-159) is an important source of income for the county government. One Tibetan from the area says that the income from the factory far exceeds the financial contribution of local taxes. According to several reports the factory will also pay the wages of government staff if the county authorities are unable to pay them. There have been several unconfirmed reports of action taken by the villagers in opposition to the factory, including one report that villagers blocked the road from Rebgong to Xining in 2001 in protest. Other reports state that local people have made representations to the county and prefectural governments and have even sent a petition to Beijing, voicing their concerns over pollution from the factory. In each case, they were reportedly told by local officials that the county depends on the factory and that the factory is the main source of income for the salaries of the staff of the county government.

54 Township Village Enterprises (TVEs), the establishment of which led to much of the wealth creation on China's eastern seaboard.

The cost of funding environmental protection can also be prohibitive, particularly for smaller enterprises – spending on environmental protection cuts into these profits. Even the larger State Owned Enterprises (SOEs), required by law to establish environmental units or staff,[55] are now facing increasing competition and are having to improve efficiency and profit ratios. As a result of deregulation, failing enterprises can no longer rely on the state to bail them out. There are also considerable personal incentives for key officials in each area to be entrepreneurial and maximise economic growth in the territory under their jurisdiction.[56] This has been apparent in Tibetan areas, as throughout China, where widespread local protectionism and corruption has caused problems for both local populations and the central authorities.[57]

In their attempts to meet what can often be contradictory demands from higher levels – and in some cases attempts to make personal profit at the same time – officials will often be less than truthful when reporting to higher levels. This makes it difficult to assess the true extent of environmental degradation and certainly means that it takes longer for serious problems to come to light. This phenomenon of false reporting has parallels throughout the history of the PRC. For example, during the 1950s, official pressure to meet unrealistic targets meant that inflated statistics were issued on grain production, resulting in widespread famine following the 'Great Leap Forward'. Although a great deal has changed in China since the 1950s, local officials still strive to protect their own interests, whether they are political or economic. Unless incentives to protect the environment are increased, either through positive or negative means (rewards or sanctions), environmental protection is not going to be accorded a high priority in the face of pressing economic interests.

China does have an established administrative structure separate from local government that is responsible for monitoring environmental protection. The State Environmental Protection Agency (SEPA) comes directly under the State Council and operates at ministerial level. There are Environmental Protection Bureaux (EPB) at each level of government from provincial to township. However, Kenneth Lieberthal, a professor at Columbia University, states that the way in which the administrative hierarchy is structured in China limits the effectiveness of these organs. He explains that priority is generally accorded to the needs of the locality vis-à-vis EPBs of the same level:

55 Yohei Harashima, 2000, op cit.

56 Kenneth Lieberthal, 'China's governing system and its impact on environmental policy implementation', China Environment Series Issue 1, fall 1997, op cit.

57 See Chapter 3 for examples of the influence local officials can have over mining projects in Tibetan areas and in overriding local concerns. See also Chapter 2 for a wider discussion of the difficulties that the central authorities have faced in controlling and regulating resource exploitation.

> *"This structure of authority effectively puts each EPB under the thumb of precisely the officials who have the greatest responsibility for – and interest in – accelerated development of the local economy. Environmental policy implementation inevitably suffers accordingly".*[58]

China appears to work according to the 'polluter-pays' principle, summed up by the Qinghai Legal News as "*Whoever develops, protects; whoever pollutes, cleans; whoever damages, repairs*".[59] However, enforcement of this principle involves a degree of supervision (and political will) that is lacking at present. The Tibetan who ran a small private mine in Lithang county, Kardze TAP, said that in his view refilling mine pits was not the responsibility of the gold miners:

> *"This is the responsibility of the Chinese government, because they are the ones who sell the land – the nomads are not allowed to do so. When mining started the Chinese government told the nomads that the government would restore the place after mining stopped, and they gave some money to the nomads. Then the government sold us the land."*[60]

This miner was clearly unwilling to take responsibility. Whether this was through ignorance of regulations or for economic reasons (or perhaps a combination of both), it is clear that nothing was done on a local level to enforce restoration of the grasslands. Private gold mining was reportedly banned in the county in 1999 when a new county head was appointed. According to unconfirmed reports the ban was linked to a scandal involving the previous county head, who was transferred after it was discovered that he had been illegally profiting from gold mining in the area.

The Chinese media has paid considerable attention to government crackdowns on small-scale mining. However, other policies that have been introduced seem to contradict the purpose of these crackdowns. For example, the TAR introduced new mineral regulations, effective 1 July 1999, that were to "*encourage farmers and herdsmen to develop mines to promote economic development*", whilst also stipulating that responsibility for protecting resources and the ecological environment lies with those who are developing the mines.[61] The environmental consequences of this type of private small-scale mining when it was carried out in Lithang in previous years were clearly devastating (see above).

58 Kenneth Lieberthal, op cit. He states that there are cases in China where a local EPB will impose a fine on a large local enterprise; the fee is passed into the government coffers. The government then provides a tax break to the enterprise, roughly equivalent to the fine that was originally levied. In this way the EPB meets its responsibilities by imposing the fine; the government meets its responsibilities by maintaining the financial health of an important source of local jobs and income. Nothing is achieved to protect the environment.

59 'Qinghai has regulations for developing mineral mining for prefectures', Qinghai Legal News, 16 January 2002

60 TINDoc 446(jw). Although he refers to the land being 'sold' by the authorities, it will in fact have been leased.

61 'Mineral rules aim to encourage farmers, herdsmen to develop mines', Xinhua, 3 June 1999; SWB FE/3553, 5 June 1999

There is little incentive for small-scale private miners, even if they are operating legally, to employ high environmental protection standards. It is inevitable that individuals running these mines will look to maximise short-term profit, leading to extensive rather than intensive mining and considerable waste of resources as well as a general disregard for any measures to protect the environment.[62] Official regulation of artisanal and small-scale mining is also influenced by economic interest. It is not only official mining enterprises (state and private) that provide important economic income to local governments and individual officials. Local authorities have also benefited from illegal and quasi-legal mining.[63]

Even where local officials do attempt to control mining activities, the pressures from both miners and other interested parties can be very strong. For example, one source reported that the deputy head of Nagchu prefecture had made attempts to clamp down on gold mining because of the damage being done to the pasturelands. This was reportedly met with firm opposition and threats from prospectors.[64]

Education/consultation

Some commentators point to a lack of awareness of environmental issues and insufficient education amongst both officials and the general population. Many of the environmental projects being carried out throughout China have focused on the need to generate public awareness of environmental issues. As illustrated by the example of media reportage on dam construction in China, public education and awareness-raising on environmental issues remains confined within certain parameters. While public education will certainly be an important factor in the improvement of environmental protection in China, it will be limited as long as the central authorities are unwilling to address the environmental impact of central policy. The Chinese authorities continue to show a tendency to blame China's herders and farmers for environmental problems, without taking into consideration the role that its policies have had in creating these problems. For example, a SEPA official, Yang Chaofei, commenting on a 2001 SEPA survey of the environment of the western regions said that the main cause of environmental destruction is poverty and that local people "*should take the bulk of the responsibility*".[65] According to Xinhua the survey reported that:

62 Or, as was discussed in the previous chapter, the health and safety of workers.
63 See p.171-179, for further discussion of this phenomenon in the case of Chumarleb county, Yushu TAP.
64 TINDoc 40(rp)
65 'Study shows environment in western region worsening', Xinhua, 29 December 2001

> *"Local people destroyed large tracts of forest and grassland to make way for grain production without considering whether the land was suitable for cultivation. [...] Massive construction of dams and irrigation projects in land-locked areas may cause water supply shortages in the lower reaches of rivers, and shrinking of lakes, oases and wetland."*[66]

Population pressure and a low level of environmental awareness amongst local populations undoubtedly have a negative impact on the environment. However, SEPA appears to ignore the fact that previous state policy required the conversion of forest and grassland into agricultural land. For example, much of the land reclamation in Tsonub M&TAP, Qinghai, was carried out by prisoners and forced immigrants in the *laogai* network and agricultural settlements developed during the 1950s to support mining and other industrial development in the prefecture. Likewise, it is the state, not local people, that has been the driving force behind dam building in China, including the Three Gorges dam (and irrigation projects, particularly large scale initiatives). One of China's main priorities for its Tenth Five-Year Plan is the construction of key hydropower projects in western China, including Tibetan areas rich in water resources, to power resource exploitation and to transfer electricity from west to east China.

If education on environmental protection were to be combined with consultation over mining and other development, this could prove more effective. In Tibetan areas at least, much of the education or propaganda that comes from higher levels is distrusted to a large extent by its intended audience. The credibility of environmental propaganda can be seriously undermined when local people witness continued damage as a result of mining or other government initiatives (from which they gain little or no benefit) and the corrupt practices of local officials.

66 Ibid.

Gold rush in Qinghai – case study

Gold mining appears to have had a wider impact on the Tibetan environment than other mineral resource exploitation,[67] due to the fact that gold is relatively cheap and easy to extract and process, potentially providing a quick profit. This has led to widespread localised small-scale exploitation by both authorities and private prospectors.

The environmental risks associated with mining have been exacerbated by poor protection in small-scale gold mining, where the short-term interests of stakeholders overrule the longer-term environmental concerns; and where inadequate measures are in place to monitor or regulate standards. Gold mining can be particularly damaging due to extensive excavations, resulting in vegetation loss and soil erosion through loss of fragile topsoil; and the use of cyanide and/or mercury in the separation of gold. Both of these factors, when combined with poor regulation and management and insufficient clean-up operations can have a catastrophic effect on the local environment and threaten the livelihoods of local communities who depend on the land and water.

A gold rush, encouraged by Chinese press reports of the riches to be found on the gold fields of the plateau, hit Tibetan areas particularly severely in the early to mid-1990s, causing a massive influx of prospectors and migrant workers. The areas worst affected were Nagchu prefecture in TAR,[68] Kardze prefecture in Sichuan and several areas of particular ecological significance in Yushu TAP, Qinghai, including the Kekexili (Mong: Hoh Xil) nature reserve and the 'Three River's Sources District (*san jiang yuan diqu*)' area, covering Chumarleb, Tridu (Ch: Chengduo) and Dritoe (Ch: Zhiduo) counties. The three rivers, of major ecological importance to China and Southeast Asia, are the Mekong (Tib: Dzachu; Ch: Lancang), the Yangtse (Tib: Drichu; Ch: Changjiang) and the Yellow River (Tib: Machu; Ch: Huanghe).

While there have been several Chinese press reports on official attempts to clamp down on illegal prospecting and to clear up mining areas in Qinghai during the last decade, subsequent reports have indicated that the problems have continued. In February, Qinghai Daily reported that there were still 20,000 people a year entering Qinghai's gold fields to carry out illegal prospecting.[69]

67 This does not include the large-scale deforestation that has occurred in Tibetan areas since the PRC was founded, particularly in Kham, now largely incorporated into eastern TAR (Chamdo prefecture) and western Sichuan (Ngaba and Kardze TAPs).
68 Much of Nagchu prefecture lies within the boundaries of the Jangthang (Ch: Changtang) nature reserve, established in 1993
69 'Qinghai province bans gold digging to protect the environment', Qinghai Daily (in Chinese), 9 February 2002

In addition to the impact of illegal 'gold rush' mining, economic reforms, particularly since 1992, have placed greater pressure on local authorities to push forward economic development and open up resource exploitation. As a result, during the 1990s gold mining became an important source of local (county-level and above) government income in some Tibetan areas. This has meant that as well as establishing their own small-scale gold mining enterprises, as county or township enterprises, (generally reported to employ poor environmental standards), local governments have also profited from semi-legal and illegal mining by selling mining permits, buying gold from prospectors and unofficial trading.[70]

Gold mining in riverbed, Kardze TAP **© Anders Højmark Andersen, 1994**

70 According to a report from the Lanzhou Morning News, local authorities do not always benefit from the illegal mining of gold reserves in their boundaries. Exploration by a People's Armed Police (PAP) gold unit in 1998 found a 17km gold belt which they named the Yangshan gold mine, running through Wen county, Gansu and reaching into Jiuzhaigou county in Ngaba Q&TAP, Sichuan, at its western end. Local villagers and others from outside started illegal gold mining, or renting out their land to prospectors (again illegally). The prospectors worked digging out rock and then using makeshift panning beds from bamboo mats lined with plastic, using water and sodium cyanide to separate the gold from the rock. The landscape was scarred and surface and ground water was contaminated.

According to this report (10 January 2002), Wen county's deputy Party secretary Chen Langxing said that the county was facing serious financial problems and that far from adding to county income, the gold resources had become a serious drain. The province had not yet allowed official exploitation of the gold, but the county was responsible for protecting the reserves (including the closure of 410 illegal mining areas and destruction of more than 400 cyanide pits in 2000) and for resettling farmers, costing them considerable financial resources. According to Chen, "*the county is having difficulties in functioning*". ("*China's Number One Gold Mine Suffers Gold Rush*", Lanzhou Morning News, 10 January 2002)

By 2002, it was apparent that the economic and environmental costs of both official and unofficial gold mining in Qinghai province were finally being taken seriously at higher levels where a formal commitment to dealing with the problem has gained some ground over the economic interests of lower levels of authority. In February 2002, the Qinghai provincial government issued a "*Notice Forbidding the Extraction of Alluvial Gold within the Province*".[71] This means that the authorities have decided to ban all official alluvial gold mining, as well as renewing efforts to clamp down on illegal mining. The only exceptions to this ban are two mines in Pema (Ch: Banma) county in Golog (Ch: Guoluo) TAP and the Zhaduo gold mine in Tridu county, Yushu TAP, reported to be one of the largest in the province.[72] These three mines will be permitted to operate if they take measures to protect the environment, for example by refilling pits.[73]

A Qinghai Daily report on the ban acknowledged the financial impact that it may have on local governments and explained the reasons behind the ban, emphasizing environmental damage:

> *"Although gold digging activities have brought in income for the local governments over a certain amount of time, as exploitation has taken place over many years there has been large-scale damage to river-beds, flood plains and so on, bringing about serious damage to the natural environment and to plant life in gold mining areas. Most of these areas are nature protection areas, including the Three Rivers' Sources [district] and the Kekexili [reserve], where the fragile environment, once damaged, is very difficult to restore".*[74]

One of the areas worst hit by the gold rush was Chumarleb county, Yushu TAP. The environmental consequences have included serious damage to pastureland, decimation of wildlife, including many endangered species and the contamination of water sources from cyanide and mercury used by miners. The fact that the headwaters of three of Asia's major rivers are located in this area of Yushu adds to the seriousness of such environmental damage.

71 Reported in Qinghai Daily on 9 February 2002 as having been passed "in recent days [*jin ri*]" (ibid).
72 A previous Qinghai Daily reported stated that the Zhaduo gold mine was the "*biggest natural resource exploitation project in Yushu*" ('Exploring the 'Corridor of Gold'", Qinghai Daily, 18 January 1999).
73 Qinghai Daily, 9 February 2002
74 Qinghai Daily, 9 February 2002

The history and scale of the gold rush were described in a Qinghai Daily article dated 18 January 1999:

> *"In 1982, several old gold peasants [lao jin nong] from the time of Ma Bufang*[75] *led several thousand new gold peasants into Chumarleb county to dig for gold, and they made a rich haul. One brought ten and ten brought a hundred, as the eastern agricultural area became flooded with thousands and thousands of gold peasants entering Chumarleb county [...] At the main gold fields the number of people exceeded 30,000 at the busiest times. Along the more than 180km long and 2-3000m wide Yazha riverbed from Jiudao ravine in Zhajia pastoral committee to Yege village, the gold diggers were thick on the ground, conducting 'carpet-style mopping up [di tan shi de da sao dang]'."*[76]

The registered population of Chumarleb at this time was about 20,000. China's Legal Daily reported in 1991 on 'gunfights and rioting' amongst prospectors, 'tens of thousands' of whom had arrived in the area following the discovery by local farmers the previous year of a single nugget of gold weighing more than seven kg. The newspaper also reported that water supplies had been poisoned by mercury and cyanide and that gold diggers had 'desecrated sacred sites of Tibetan Buddhism'. Over 35 people were reported to have died between 1990 and 1991 as a result of battles between rival groups, run by 'brutal gang-leaders'. Another 20 miners were reported to have died from disease and more from falling to their deaths while smuggling gold through mountain passes.[77]

One of the main problems for the Chumarleb authorities was a lack of manpower to supervise prospectors or provide adequate law enforcement. A Western ecologist who visited Chumarleb in 1993 talked about the lack of control over migrant miners:

75 Hui warlord and governor of Qinghai 1936-1949. When the Qing dynasty fell in 1911, a regional Hui clan, the Ma, usurped power in the former Qing administration in Xining and attempted to extend political and economic control into the Tibetan territories beyond. Although Qinghai Province was formally founded in 1929, the area outside Xining-Haidong remained in a state of virtual chaos throughout the Republican period. Tibetans and Mongols answered the Ma regime's aggressive penetration of their territories with armed resistance. (Susette Cooke, 'The politics of population transfer: an ethnographic and historical survey of Haidong and Haixi', TIN Special Report, 28 October 1999)

76 'Exploring the 'Corridor of Gold", Qinghai Daily, 18 January 1999

77 Legal Daily, cited in 'Chinese gold-rush fever triggers violence', The Independent, 8 October 1991

> *"We saw between 200 and 500 men prospecting for gold, some with machetes, some with machine guns. There are about 40 police, so they cannot really control them. We were told in a meeting with local officials that Chumarleb was one of the main places where they go, and at the end they agreed that the gold miners were a big problem for the local wildlife. They stay for a month, they come with pipes and pans for panning, using simple systems – no power tools. It is a kind of poverty alleviation method, a way of providing income for rural poor from elsewhere."*[78]

The lack of human resources will certainly have prevented the authorities from controlling the behaviour of prospectors, including poaching and fighting. However, due to the economic interests of local authorities, it is far from clear that the prospectors were ever discouraged from actually mining gold. A report in the Qinghai Daily states that during the gold rush period "*a large part of Chumarleb county's income came from the gold fields*", leading to the obvious conclusion that the local authorities had been making money from illegal gold mining. As a result, when the prefectural authorities decided to close off the gold fields in 1993, the county (with prefectural approval) started leasing out enterprises to individuals for mechanised extraction to make up for the lost income.[79] These enterprises were required to refill earth and plant grass according to the contract that they signed.

The decision by Yushu prefecture to clamp down on illegal mining in the province in 1993, coincided with central government attempts to regain control over the gold industry and is likely to have been part of this wider policy development.[80] During the early 1990s, the central authorities had become increasingly aware of the extent of illegal gold mining throughout the country and the involvement of local authorities in this. Much of the illegal extraction and trade was being operated or at least overseen by local officials or the military who would then benefit from sales of gold or from charging fees for access to mining areas. This level of unofficial exploitation was damaging to the economic interests of the state.[81]

78 See also Chapter 4 on Tibetans who have participated in gold mining as a 'side job' to boost their income.

79 Qinghai Daily, 18 January 1999, op cit.

80 Another region that the government clamped down on during this period was the Xiaoqinling mountains of Henan and Shaanxi province, where an estimated 40,000 illegal miners were at work by 1992. The illegal mining effected the production of state mines operating in the same area, led to serious waste of resources as recovery rates were less than 50 per cent, and inevitably led to huge environmental damage, including pollution of water sources with mercury and cyanide. Following two years of little success, Beijing applied high-level pressure, threatening the provincial heads of Henan, Shaanxi and Shanxi (believed to be the main source of the prospectors) with losing their jobs. The area was finally cleared out by the PAP and checkpoints were set up. (Bruce Gilley, 'Golden Crusades: Beijing wins a battle but is losing the war with illegal miners', Far Eastern Economic Review, Vol. 160, No.47, 20 November 1997)

Najartse mine near Gyantse (Ch: Jiangzi), Shigatse (Ch: Rigaze) prefecture, TAR **© TIN 1996**

According to the Chinese press, the 1993 prefectural decision to clamp down on illegal gold mining was tackled by bringing officials and security personnel into the area and establishing checkpoints on main roads to block off the gold fields.[82] However, enforcement was clearly weak as illegal mining continued. In 1995, UK national newspaper The Guardian published two articles on continued lawlessness in the region. One of the reports quoted an ex-gold boss as saying:

> *"I've heard the bullets whizzing past my ears and seen people beside me dying. [...] Even if the government wanted to control us, we are stronger than they are right now."*[83]

81 By 1994, the Chinese government estimated that there were 240,000 people involved in illegal mining in China, with an estimated output of 12 tonnes of gold that year. (Far Eastern Economic Review, 20 November 1997, op cit.). When compared with an official gold production in 1994 of 90 tonnes, the 12 tonnes of gold illegally produced represented a significant loss for the state who held the monopoly on the purchase and sale of gold. Added to this was the loss of future economic potential due to the massive waste of resources from reckless mining. This led to attempts to increase controls over the gold industry and a movement towards reform. As well as attempts to regain central control over the industry for economic reasons, the authorities were also becoming increasingly aware of the huge environmental damage that illegal mining activities had caused. See Chapter 2, pp. 59-63 for more information on the development and reform of China's gold industry.
82 Qinghai Daily, 18 January 1999, op cit.
83 The Guardian, 31 July 1995

The reasons for this failure to implement restrictions on gold mining in Chumarleb were probably a continuation of lack of funding and resources to carry out the crackdown in a meaningful way and insufficient political will due to continued local interest in mining activities. The mechanised mines, although meant to limit environmental damage, have nevertheless added to the environmental degradation of the area and do not appear to have been well regulated. A report on 25 April 2002 indicates that despite official reports on the various steps taken to control gold mining, the problems had not been dealt with:

> *"In recent years, although management has been strengthened and mechanised extraction has been undertaken, due to the long-term reckless and wanton excavation environmental degradation has continued to worsen and there are criss-crossing pits and trenches and an accumulation of mountain-like spoil heaps. Putting the environment in order in mining areas is extremely urgent."*[84]

An earlier Qinghai Daily article (28 December 2001) on the Three Rivers' Source District attributed continued environmental destruction to both continued illegal and official mining. The article states: "*Due to the driving force of economic benefits, the phenomenon of reckless and indiscriminate mining has repeatedly been banned but has not stopped*". It refers both to "*illegal gold mining activities*" and also "*those operating beyond their limits*", indicating that approved enterprises are mining outside their designated areas.[85]

The Chinese authorities have expressed particular concern over environmental damage in the Three Rivers' Source District. In fact, before the ban on alluvial gold mining was instituted across the whole province in February 2002, the authorities in Yushu had already announced that all mining activities in this area had been shifted to filling in all pits.[86]

Since the ban, county level 'legal inspection' teams have been set up by the Province and given legal and semi-military training, patrols have been formed in some areas to police the gold fields, a propaganda campaign has been launched to educate both officials and the public, and in some areas working groups have been dispatched to close mines and clear up gold fields and refill pits. The various press reports on the campaign to shut down gold mining in Qinghai give the impression that this move has more momentum now, particularly given the commitment to stop official as well as illegal mining. With the weight of the provincial authorities behind it, the crackdown has a better chance of success than previous lower level initiatives.

84 'Chumarleb bans gold digging, fences land and cultivates grass', Qinghai Daily, 25 April 2002
85 'Three Rivers' Source District rigorously checks panning and digging; all pits completely refilled', Qinghai Daily, 28 December 2001
86 Ibid.

However, the extent to which the ban will be adhered to remains to be seen. Implementation will be dependent upon the importance accorded to it by higher levels[87] and will require strong regulation and monitoring. For example, even though checkpoints are reported to have been set up on the roads heading into the gold fields, there is still a risk that checkpoint guards will be persuaded by a cash payment to turn a blind eye to prospectors wishing to pass through. Local authorities will lose an important source of income; they are also likely to be expected to meet at least some of the costs of clean up operations and guarding/patrolling of reserve areas.[88]

Past experience shows that provincial dictates do not guarantee implementation. Likewise positive press coverage of environmental protection initiatives can sometimes be misleading. For example, in 1995 the Qinghai provincial government announced a decision to close all gold mines in the Hoh Xil nature reserve from 1 January 1996 to protect the region's gold reserves and wildlife from the estimated 30,000 farmers entering illegally each year.[89] This was not entirely successful as, on 30 December 1999, China Daily reported that the provincial government had (again) ordered closure of the reserve after more than 1,000 people were suspected to have entered the reserve illegally in 1999 to mine gold or hunt animals.[90]

In addition to the ban on alluvial gold mining, there were reports of crackdowns on other mining enterprises in Qinghai. In a general report on Qinghai "*comprehensively rectifying the order of the mining industry*" over the past few years, Qinghai Daily reported in June 2002 that the provincial land and resources department has launched a series of propaganda exercises "*repeatedly publicising the importance of treasuring and protecting mineral resources*". The report stated that the authorities had clamped down on salt and coal mining, as well as gold mining:

> *"[The department] has completely cleared the illegal salt fields from Cha'erhan Salt Lake, stopped production and rectified 11 production lines at Mahai Salt Lake and removed their equipment and installations from the mining area, and there has been a thorough clearing up of illegal gold mining in places including Xiao Gan'gou [little dry gully] in*

87 For example, the logging ban in Tibetan areas, instituted following the Yangtse flood disasters, has been largely effective. It was a high priority of the central authorities due to the impact that logging had had on the rest of China.

88 Chumarleb county is reported to have spent 740,000 yuan on what appear to have been largely unsuccessful initiatives to clear its gold fields in 1993. (Qinghai Daily, 18 January 1999, op cit).

89 'Qinghai's Hoh Xil region to close all gold mines in January to curb illegal mining', Xinhua, 14 October 1995; SWB FE/2436, 17 October 1995. According to the report: "*State-owned and collectively-owned gold mines with modern equipment and technology are encouraged to develop gold mines in the region in the future on the condition that gold production be planned and grassland in the region protected.*"

90 'Qinghai closes nature reserve', China Daily, 30 December 1999. This does at least represent some improvement over the reported 30,000 farmers coming to the region annually by 1984 to mine and hunt animals (Xinhua, 14 October 1995, op cit). Problems with enforcing regulations exist throughout China. For example, repeated instructions from central and provincial governments for small coal mines to close down until pollution and health and safety targets were met, have been widely flouted.

> *Yushu and Golog. Up to the present, three illegal salt lake mines have been closed according to the law, one enterprise was closed for not having a permit and two have been ordered to cease production. Five gold digging areas have been cleared, and 109 small unlicensed mining boats have been eradicated. After rectification, 50 coal mining extraction permits were cancelled throughout the province, 27 mining permits were suspended, 37 mining permits were recalled and six mine shafts were blown up in accordance with the law."*[91]

While the authorities in Qinghai province have gone further than ever before in their stated intentions to tighten controls over mining, there is clearly still a desire to exploit the province's mineral resources. Four days before the Qinghai Daily announced the steps Chumarleb was taking to clamp down on illegal gold mining it had announced the discovery of an "*exceptionally large gold mine*" with reserves exceeding 100 tonnes, in the Bayan Gala mountains. The mine is located near the source of the Yellow River and within the boundaries of Chumarleb county. The press coverage of the discovery of new deposits alongside a clampdown on small-scale mining could be a sign of the shift in policy towards large-scale state extraction, being made increasingly possible by improved infrastructure.[92]

The publication of a report in January 2002 on 33 bedrock gold deposits in the western Qinling Belt, 16 of which are located in Kanlho, Ngaba and Kardze TAPs in Gansu and Sichuan,[93] points to possible future environmental concerns over large-scale extraction of gold deposits. According to an exploration geo-scientist who has studied the report (published in the academic journal Mineralium Deposita in January 2002), the presence of toxic elements, particularly arsenic and mercury, that are commonly associated with these gold deposits, means that particular care is required when separating the gold from the ore to avoid the leakage of toxic elements into any water system. Another concern is that these deposits are likely to be mined using the cyanide heap-leach process, which is relatively low cost. It is not yet clear whether it will prove economic to mine these deposits. However, if they are developed under current environmental safety standards there is a clear risk of leakage of cyanide solution and other toxic elements into water systems, as has apparently already occurred in other mines.

91 'Qinghai province rectifies the order of the mining industry', Qinghai Daily, 6 June 2002

92 See Chapters 1 and 2

93 Jingwen Mao, Yumin Qiu, Richard J. Goldfarb, Zhaochong Zhang, Steve Garwin, and Ren Fengshou, 'Geology, distribution, and classification of gold deposits in the western Qinling belt, central China', Mineralium Deposita, 29 January 2002. See Chapter 1 p.27

Top: Gold mining near Lake Manasarovar, 2002
Below: Gold-mine drainage into Lake Manasarovar, 2002

6. Costs and benefits

The Chinese authorities are quite open about their aims to open up western China to large-scale resource exploitation in order to support domestic industry and remain as self-sufficient as possible in the supply of minerals. However, at the same time they emphasise that this will lead to the enrichment of the 'minority nationality' peoples living in these areas; two of the main themes of official literature on the current drive to develop the western regions of China are poverty alleviation and environmental protection.

The industrial development model that the Chinese authorities are pursuing in Tibetan areas, based on the exploitation of mineral resources, is seen by many economists as an engine of development in poor areas well-endowed with minerals. However, others challenge this view on the basis of the experience of countries or regions where the people have not benefited and mining has actually enhanced economic and social disparities, as well as causing long-lasting environmental damage.

Some mainstream academics within China are now arguing that industrialisation based on mineral resource exploitation is not appropriate for Tibet for economic, social and environmental reasons, and are calling for the adoption of an alternative ecologically and culturally appropriate development model. This chapter discusses some of these views and draws on the documentation of previous chapters to assess the extent to which Tibetan areas and Tibetan people have or will benefit from resource exploitation.

The first section of this chapter examines the sustainability of the current model of development in Tibetan areas, questioning the idea that the exploitation of Tibet's rich natural resources is necessarily a precursor to the enrichment of local or regional economies. It also challenges the principle that resource exploitation will benefit Tibetans by alleviating poverty and enhancing income generation and looks at the economic sustainability of the current development model. This section notes that Tibet's fragile environment is one of the main arguments against the current development model, although environmental issues are primarily dealt with in Chapter 5.

The second section of this chapter looks at the distribution of benefits of mining in Tibetan areas, providing a general outline of the distribution of revenue on a regional, local and community level. It also looks at the balance of costs and benefits of mining to Tibetans, discussing the lack of control that Tibetan areas and Tibetan people have over their economic futures and the likelihood that current development policy will serve to increase economic, social and cultural marginalisation. This section also mentions the potentially disruptive impact of mine closure on communities that are linked to the mining industry if long-term development plans are not put in place.

The third section brings the discussion into the framework of the international debate on sustainable and rights-based development, outlining the basic principles of accountability and participation and what 'sustainable development' should mean for a local community. It then looks at the difficulties in implementing these principles both internationally, but also specifically in Tibet given the current lack of mechanisms in China to ensure accountability, transparency and genuine participatory development. The final pages outline the politics of development in Tibetan areas and raise the question of whether or not it is preferable for international agencies to be involved in China's plans to mine Tibet.

Model of development

Rich resources equal wealth?

Minerals are often among the few promising economic resources in developing countries, but their exploitation has frequently distorted economies and local people tend to lose out to the interests of national authorities. The economic growth (as measured by GDP) of countries that have industrialised on the basis of mineral resource exploitation is generally much slower than countries with poor resources but an abundance of manpower.[1] There have been many reasons for this. Raw materials are worth considerably less than processed minerals or manufactured goods. As a result, countries or regions that export raw materials benefit less from their resources than those importing raw materials and producing semi-finished and finished value-added products. In addition, the international minerals market is susceptible to fluctuations and in the current international environment minerals producers are facing increasing competition.[2] The environmental impact of mining can prove costly in the long-term for mineral producing countries or regions. Other associated factors that have limited wealth generation through mining include poor governance and corruption and mineral exploitation in many resource rich countries has been associated with 'bad government'.[3]

1 Asian Development Bank, 1997, cited in Hu Angang and Wen Jun, 2001

2 See Chapter 2, p. 66

3 For example, according to Human Rights Watch, during the late 1990s the Angolan government used oil revenues to buy covert arms that undermined the spirit of peace accords that had been agreed in order to end the civil war in which hundreds of thousands of civilians had lost their lives. The covert arms purchases bypassed the Central Bank and the Ministry of Finance and were routed instead through the state oil company (Sonangol) and the Office of the Presidency. 'The IMF and Angola: Oils and Human Rights', Human Rights Watch press release, June 2000

Nigeria is a good example of a country that is rich in mineral resources, but has a very poor population. The development of the oil industry in Nigeria has occurred at the expense of other potential development, leaving a narrowly focused economy dependent on oil.[4] The benefits from oil have been spread very thinly among the country's governing military elite, while local people have received little or no benefit and have had to live with the environmental damage caused by oil exploitation. Although the World Bank is in support of poverty alleviation through mineral exploitation, it acknowledges that living in a resource rich country does not necessarily make a population rich:

> *"Despite the country's relative oil wealth, poverty is widespread and Nigeria's basic social indicators place it among the 20 poorest countries in the world... Economic mismanagement, corruption, and excessive dependence on oil have been the main reasons for the poor economic performance and rising poverty."*[5]

The distribution of resources and manpower relative to wealth generation in China reflects a global trend. While mineral resources are concentrated in the remoter and sparsely populated regions of China, the eastern seaboard, with an abundance of manpower, has experienced much faster economic growth. Chinese state policy remains based on the principle famously outlined by Zhou Enlai in 1957 of 'mutual assistance' among the nationalities of China to overcome the imbalance of resources and manpower. This involves building up human resources in the resource-rich west to enable the exploitation and transfer of raw materials to the more highly populated areas for industry, particularly the east coast.[6]

Moreover, mineral resources in Tibetan areas are not regional assets because they are state owned. As such, exploitation of raw materials does not necessarily enrich Tibetan areas. The wealth that can accrue to these areas through resource exploitation is largely limited to any potential stimulation of regional economic development through mining. As Tibetan areas have principally been providers of raw materials for the industrial development of eastern China their economies have seen little benefit.[7]

4 According to the World Council of Churches, about 90 per cent of the foreign exchange earnings of Nigeria are provided by oil – www.africaaction.org/docs97/ogon9703.htm

5 World Bank Africa, 'Countries: Nigeria', December 2001, World Bank website www.worldbank.org

6 See Chapter 3, pp. 84-94

7 See also pp. 191-193

The state, as owner of all mineral resources, has benefited directly from fees, taxes and sales of mineral products, while Chinese industry has benefited from the supply of cheap raw materials. Although some Tibetan commodities, notably gold and copper, need to be processed locally as the untreated ore contains only minute quantities of the metal, processing and manufacturing industries remain largely concentrated in the east of the country. As a result, the potential value of Tibet's minerals, in the form of value-added semi-finished or finished products, is realised by these provinces, not by Tibetan areas.

As part of Western Development, the authorities state that they are trying to redress this imbalance by developing processing industries in Tibetan areas, or by encouraging Tibetan businesses to invest in the interior. However, at present most resources continue to be transported out without value added. The development of Tibet is to be carried out under the 'unified plans' of the central authorities and according to 'market demand',[8] suggesting that economic development of Tibet remains primarily focused on production of raw materials to supply the more populous regions and profit-orientated industries of the east.

Although the state seeks to maximise the potential wealth of national resources, the determination to exploit Tibet's natural resources to reduce state reliance on importing raw materials, for example copper, will not necessarily suit the demands of the growing private sector of industry. Traditionally, the growth of Chinese industry, including the mining industry, has been production rather than profit oriented, and the Chinese authorities has striven for self-sufficiency in the supply of raw materials for industry. However, China's development towards becoming a market economy and recent entry to the WTO will lead to the opening up of domestic industry to external market influence and international competition. Producers in the east may find it cheaper and more expedient to import raw materials. This is one of the reasons that some academics within China are now arguing that Tibet should be seeking an alternative model of development and that Tibet's resources should not be exploited simply because they are there. They should only be exploited where economic, environmental and social conditions are suitable.

For example, in January 2000, Chinese academic Geng Xiangling, who worked for many years at the Tibet Minorities Institute, produced a paper entitled 'Further understanding the superiority of Tibet's natural resources'.[9] Geng Xianglin argues that the 'resource era' is now changing to an 'information era'. The traditional view of a region's resource superiority being based on natural resources is being increasingly

8 See for example Xinhua, 28 February 2001

9 Geng Xianglin (now associate Professor at Jiangsu Changshu Scientific Specialist School), 'Further understanding of the superiority of Tibet's natural resources', Tibetan Studies, 2001/1, pp.37-40.

challenged as a result of globalisation and growing international concern for environmental protection and sustainable development. This author argues that given the problems associated with exploitation of Tibet's natural resources[10] and environmental concerns,[11] *"Tibet must abandon the traditional model of industrialisation"*.[12] He states that the economic advantage of natural resources will become less and less and that Tibet should be concentrating on developing knowledge and information industries that can overcome such restrictions as remoteness, lack of transportation facilities, harsh climate and shortage of power. He also argues for the reform of traditional agriculture and animal husbandry, development of alternative energy resources (solar power, wind power, geothermal energy, nuclear energy) and non-polluting industries (medicinal products, green products, ecological agriculture/animal husbandry/forestry). Tibet's superiority, he argues, now lies in ecology, tourism, adventure sports (mountain climbing), scientific research and religious and cultural heritage. These things should be protected and developed to support economic development.

Mining as poverty alleviation?

> *"Economic development policies for Tibet adopted by the central government basically adhere to the governing tactics, advocated by traditional modernisation and development theory, that the 'fundamental way to relieve poverty lies in exploiting rich natural resources and in bringing about industrialisation'."*[13]

There appears to be increasing use of the argument amongst development agencies, mining companies and governments that exploitation of mineral resources can function as poverty alleviation for local communities. However, past mineral development has often enhanced the disparities between the rich and poor, both internationally and within nation states.

In an article published in 'China Tibetology' in January 2001, Chinese economists Hu Angang and Wen Jun tackled the issue of Tibetan modernisation, and specifically the designation of mineral exploitation as the basis of Tibetan modernisation.[14]

10 see Chapter 2
11 see Chapter 5
12 In this paper, the author is referring specifically to the TAR when he talks about 'Tibet' (Ch: Xizang). However, his argument stands for other Tibetan areas and has been echoed by other researchers and analysts who consider that this model of development is not appropriate for the Tibetan plateau.
13 Hu Angang and Wen Jun, 'The Problem of Selecting the Correct Path for Tibetan Modernisation (Part 1)', China Tibetology, 2001/1, pp.3-26. Hu Angang is a professor at the 'Development Research Academy for the 21st Century' at Qinghua (Tsinghua) University, Beijing. Since 2000, he has been a commissioner on the 'Experts Commission on Territory Resources' of the Chinese Academy of Sciences. Wen Jun (Ph.D) is based at the Public Management College of Qinghua (Tsinghua) University
14 Ibid.

The authors challenge the traditionalist modernisation economic theory that considers industrialisation based on natural resource exploitation to be the way to promote regional economic development and to relieve poverty. They argue instead that agriculture (including animal husbandry) should be the basis of development in Tibet.

According to Hu and Wen, problems in China's economic development originated in the early 1950s when the strategy of giving precedence to heavy industry was adopted, attempted by policies to 'squeeze agriculture for industry [*yi gong ji nong*]', or in other words using surplus agricultural production to support rapid urbanisation and industrialisation. Hu and Wen state that since reform and opening up, the direction of policy moved towards equalising agriculture and industry, so that agriculture no longer offers up its surplus production for the development of industry and each sector is self-reliant, exchanging goods of equal value. However, in the past few years the TAR has reverted to a strategy of sacrificing agriculture to industrial development. Hu and Wen argue that for regional development, agriculture and animal husbandry are the most important industries. They also provide statistics in support of their argument that they are also the most productive and profitable industries. In 2000, Hu and Wen point out, 86 per cent of the TAR's population still lived in the rural areas.[15] They stress that regional development should be focused on the majority population:

> *"The choice of the road to modernisation must consistently uphold the basic principle of 'enriching the people as fundamental, investing in the people [fu min wei ben, tou zi yu ren min]', to enable the people who constitute the absolute majority of the population – the peasant farmers and herdsmen – to become first-priority, direct and universal beneficiaries, as quickly as possible."*[16]

According to the official press no contradiction exists between the economists' call for enrichment of the people and current Chinese policy. China's official position is that development of the mining industry will stimulate the local economy of Tibetan areas and will contribute to poverty alleviation. According to the China Daily: "*Mining development will stimulate the growth of local township enterprises, absorb redundant labour in rural areas and increase the income of Tibetan people*". In an article in Tibet Daily in 1996, the general manager of the Nagchu Prefecture Mining Development Company talked about the problems of enriching the 'livestock rearing masses', stating that relying on animal husbandry has not allowed overall economic development.

15 This figure accords with official population statistics for the TAR population in 2000 as listed in the TAR Statistical Yearbook, 2000, and will not include the floating population of urban areas.
16 Hu and Wen, 2001, op cit.

Mining, he says, will:

> *"increase wealth in the region, stimulate rapid development of interrelated industries, absorb surplus labour power in the pastoral areas and raise the material and cultural lives of the pastoral masses, who could begin to escape poverty and attain wealth quickly".*[17]

However, so far the development of industry (based on resource extraction) has resulted in an increasing gap between rich and poor. The majority of Tibetans are still farmers and nomads and it is this sector of the population that is widely acknowledged to have received the least benefit from development so far. Hu and Wen argue that the current model of development is not an effective way of lifting the majority of Tibet's population out of poverty, and will actually slow down the economic development of the TAR:

> *"Not only will [the traditional policies of industrialisation] be unable to truly give impetus to modernisation and development in Tibet, but also, it will be really difficult for the peasant farmers and herdsmen who are the main grouping in society to benefit from them. Not only will the peasant farmers and herdsmen lack opportunities [wu yuan] in the course of modernisation and development, they could also become an extremely serious burden in the economic development of Tibet."*[18]

That farmers and herdsmen have not been the principal beneficiaries of economic development was also acknowledged in an official analysis of China's 'Aid Tibet' projects (construction projects funded by the central government and other Chinese provinces), based on interviews with TAR residents (Tibetan and Chinese). The report, published in the 'Nationalities Research Magazine' in January 2000, focused on the success of China's two main sets of 'Aid Tibet' projects – the 43 and 62 projects – which were largely concerned with infrastructure construction (transport, communications, power) to support industrial development.[19] The author, Professor of the Social Sciences Department of the Central Party School, Jin Wei, acknowledges that while the majority of respondents were apparently in favour of China's aid

17 'The search for countermeasures to develop mining in northern Tibet [*Zangbei*]', Tibet Daily, 4 December 1996.

18 Hu and Wen, 2001, op cit. See also Wang Shaoguang and Hu Angang, 'The Political Economy of Uneven Development: The Case of China', ME Sharpe, 1999.

19 The 43 projects were announced at the Second Forum on Tibet Work (1984), when the authorities launched the 'Aid Tibet' campaign. They were medium-sized engineering projects, mainly concerned with infrastructure construction.
The 62 construction projects were announced at the Third Forum (1994) to rapidly accelerate development. They can be broadly categorised as follows: 5 agricultural (mainly irrigation); 9 industrial (including agricultural products processing); 1 forestry; 14 power (mainly hydropower); 4 mines; 9 infrastructure (roads, pipelines, telecommunications); 6 education; 5 health; 9 other (including media, government buildings, border posts, Tibet Museum and Potala Palace Square). For a full list of the projects see TIN News Review: Reports from Tibet, 1998, London: TIN, 1999, pp. 90-91

projects, more than half the farmers and nomads felt they had received little benefit from them so far. The article does not provide a figure for the percentage who felt that the projects had had no effect at all:

> *"During out investigation we asked, 'What effect have the Aid-Tibet projects had on your lives?' The respondents could select one of three responses: 'A big effect' [miqie xiangguan], 'little effect' [quanxi bu da], or 'no effect' [wu guan]. 57.95 per cent of the peasant farmers and herders selected 'little effect'."*[20]

Jin Wei states that the low impact on farmers and nomads is probably because such projects were mainly focused on urban areas. Most Chinese in Tibetan areas are concentrated in urban and industrial centres and as a result count amongst the recipients of benefits of development in these areas. This was acknowledged by Jin Wei, who states that Chinese people questioned in the research were generally more positive about the projects, in part because, as residents of cities and towns above county-level, the projects have had a closer bearing on their lives. She also states however, that the different response from Chinese and Tibetans to the projects were due to Chinese having a 'higher level of understanding', indicating that the historical tendency to discount awkward Tibetan views as 'backward' prevails.

Mining as a form of poverty alleviation actually appears to have been most prevalent in the unofficial sector of the mining industry in Tibetan areas, although often not benefiting the local community. Artisanal mining has provided extra income for tens of thousands of rural poor, largely Chinese and Hui Muslims from neighbouring provinces, but also Tibetans. If the Chinese authorities succeed in limiting artisanal and other small-scale mining and developing larger-scale exploitation, this will have a mixed effect on local Tibetan populations. For the majority of Tibetans living in or near the gold fields, the impact will be positive because at present they are left with damage to livestock and land and water sources after miners have moved on. However, those Tibetans who have engaged in gold mining as a sideline occupation, will lose the extra money that at present boosts their income from agriculture or animal husbandry.

In terms of official employment in the mining industry, there is generally a lack of local technical skills and so Tibetans have largely been limited to working in manual labour, while skilled and managerial positions go to imported personnel. As such, even if the overall wealth of an area may increase, the development of mineral resources tends to highlight and reinforce, rather than reduce, the disparities between Chinese and Tibetans.

20 Jin Wei, 'Social appraisals and evaluations of Aid Tibet projects', Ethno-National Studies (Minzu Jiangjiu), 2000/1.

Economically sustainable development?

The model of economic development that China is pursuing in Tibetan areas, based on resource exploitation and infrastructure construction, appears to be dependent on financial subsidies for completion and success. Hu Angang and Wen Jun consider that, on the basis of results achieved so far, the traditionalist model of development, at least as far as it is applied to the TAR, is of 'unsustainably low economic benefit' because it is actually increasing rather than decreasing Tibet's dependence on subsidies from the central government. The authors state that each time an industrial enterprise makes a loss of one unit, this loss is compensated by the central government 2.14 times over that value – each loss of 100 yuan (US$12) results in subsidies of 214 (US$26) yuan. By contrast, when an industrial enterprise increases its output by one unit, it receives compensation of 2.89 times – ie. increased output valuing 100 yuan, would result in a subsidy of 289 yuan (US$35). This means that enterprises are receiving greater subsidies when output increases than when they are making a loss.[21]

Jin Wei also hinted at this pattern of dependency in her report on China's 'Aid Tibet' projects. She states that factories and mining enterprises in the TAR[22] have had few beneficial results and that most of them are incurring losses. While her report found that other 'Aid Tibet' projects (for example agriculture or health) had achieved better results, she stated that most had financial difficulties and all required more investments, leading to a situation of increased reliance on the state:

> *"As [Tibet] lacks the ability to accumulate funds, a situation has arisen where the more projects the state establishes in Tibet, the greater will be the need for [state] investment to maintain and support these projects."*[23]

China has been attempting, as part of the Western Development drive, to attract investment into the mining industry in Tibetan areas. However, as commentators like Hu Angang and Geng Xiangling indicate, this is going to be difficult in the current competitive environment. The sorts of projects that the state is pushing forward in Tibetan areas, namely huge infrastructure projects and large-scale mineral exploitation,

21 Hu and Wen, op cit.
22 Four of the 62 projects were mines: the Xiangka mountain chromite mine in Chusum (Ch: Qusong) county, Lhoka (Ch: Shannan) prefecture; the Bengnazangbu gold mine in Shantsa (Ch: Shenza) county, Nagchu (Ch: Naqu) prefecture; a Szaibelyite (magnesium borate hydroxide) mine at Chagzam salt lake; and the Machala coal mine in Riwoche (Ch: Leiwuji) county, Chamdo (Ch: Changdu) prefecture.
23 Jin Wei, op cit.

need large injections of capital and technology and are very much long-term investments with no quick return on capital. Added to this is the huge risk associated with mineral exploration and exploitation. These projects may be of importance to the state for a variety of reasons – economic, political and social – but many of Tibet's minerals may remain unfeasible for commercial exploitation.[24]

A fragile environment[25]

The fragile environment of the Tibetan plateau and the far-reaching impact that Tibet's ecology has in Asia, provides one of the strongest arguments against a developmental model based on resource exploitation – particularly given the current failure to implement existing protective regulations. Despite the development of a legislative framework governing environmental protection, the mechanisms are not in place in China for implementation and good governance. While resource exploitation remains the basis of economic development, strong conflicts of interest that undermine environmental protection will continue to exist at every level.

A reflection of the importance of Tibet's environment, and the relative freedom with which environmental issues can be discussed in China, is the fact that academics both inside and outside China are calling for 'green' (ie ecologically appropriate) development of the Tibetan plateau.[26] For example, Chinese economists Hu Angang and Wen Jun state that a high proportion of the TAR's industrial output value accounted for by the mining and raw materials industry (61.9% in 1997) is not only higher than the eastern regions of China (as would be expected) but also considerably higher than other provinces and regions in western China.[27] Given past experience in China and other developing countries, they are concerned that if this strategy for industrialisation continues to be implemented, it will prove difficult to sufficiently protect the fragile ecosystem of the Tibetan plateau:

> *"Not only will this [strategy] be harmful to the protection of the environment, it could also possibly lead to history repeating itself in the form of 'pollute first, put in order later' and 'great damage, great pollution."*[28]

24 See Chapter 2

25 The environment is only briefly mentioned here as it has been dealt with in detail in Chapter 5

26 See for example, Geng Xianglin's comments on pp. 184-185

27 According to the vice-minister for Land and Resources, Shou Jiahua, the average proportion of total industrial output accounted for by the mining industry in western China is 40%, eight per cent higher than the national average. Shou Jiahua, 'Mining Development in Western China', paper given at the Round Table Conference on Foreign Investment and Mining Development in Western China, XUAR, 13 October 2000. Published on the MLR website, www.mlr.gov.cn

28 Hu and Wen, op cit.

Distribution of benefits

Distribution of revenue

The question of who benefits from resource exploitation, the national government or the locality, is a classic aspect of the international minerals debate and is particularly sensitive when local people belong to a different ethnic group and/or do not identify with the nation. As the owner of all mineral resources, including those in 'nationality autonomous' areas, the Chinese state has the rights to benefit from their exploitation and the principal revenue from mineral exploitation goes to the central government. This is by no means a situation peculiar to China. The Mining, Minerals and Sustainable Development Project (MMSD), when assessing the distribution of revenue from mining across the globe found that:

> *"Often all taxes and royalties from minerals operations have gone straight to the central government, and the only benefits that communities could expect were those that trickled down through central government spending."* [29]

Tibetan areas are designated 'autonomous', but they are nevertheless bound to implementing state plans for development. Under the present model these areas are obliged to base economic development on resource exploitation, but they do not have autonomous control over use of these resources or any rights to benefit from exploitation of minerals. Although the amended National Autonomy Law has introduced a new provision granting these areas an unspecified 'level of compensation' for resources that are exported,[30] the wealth contained in Tibetan soil does not belong to Tibetans. This means that the value of exported minerals is not attributed locally, while continued exploitation of minerals and exporting of raw materials increases reliance on subsidies for further development.

In the introduction to the CD-ROM 'Tibet Outside the TAR', authors Steven Marshall and Susette Cooke point to the fundamental issue of ownership:

29 'Breaking new ground: Mining, Minerals and Sustainable Development'. IIED. 2002, p.17. See p. 198 for more details about MMSD.
30 See chapter 3, p. 80

> *"There are no Tibetan natural resources. Trees and gold can be Chinese; they cannot be Tibetan. [...] Not only have Tibetans been constitutionally relieved of control and ownership of natural resources which they held for centuries, but they have been separated from the wealth generated by those resources. For Tibetans it is almost as if Tibetan resources never existed at all. They have become Chinese, then transferred to China."*[31]

Marshall and Cooke point out that this is doubly harmful for Tibetans since the resources over which they have the least control are those which have the greatest value and which are irreplaceable. The fact that Tibetans have no rights to control or profit from these resources prevents them from having any semblance of economic autonomy. Central exploitation of Tibet's resources as a basis for development makes these areas even more economically dependent on China and thus further integrates Tibetan areas into the greater Chinese state.

Prefecture, county and township level governments are responsible for mining medium-scale deposits and some of them have established processing enterprises that have become important sources of government income. In other cases local governments rely on revenue from leasing land out to private miners on a quasi-legal basis. However, most local income generated through the mining industry appears to be ploughed back into administration, primarily paying the wages of government workers. Very little appears to filter down to nomads, herders and farmers in the form of improved education, facilities and health-care. For example, a Tibetan who worked in a county-level administrative department in Qinghai province said that his county government collects a yearly fee for leasing land to two private gold mining enterprises. Commenting that the county does not have money to spend on such things as improvements in education, he adds that:

> *"Most of the income from the area goes towards salaries of the government staff. There is a lot of expenditure on the salaries of the government staff."*[32]

This reflects the priorities in spending in Tibetan areas throughout the current development period since the early 1990s. The proportion of budgetary expenditure on administration has been very high and in some years has exceeded all other sectors of the economy. In the TAR at least, spending on administration has alone exceeded government income.[33] This has partly been due to the large bureaucratic

31 Steven D. Marshall and Susette Cooke, 'Tibet Outside the TAR' 1997
32 TINDoc 695(jw)
33 'TIN News Review: Reports from Tibet, 1998', TIN, 1999

superstructure and also a result of the secondment of Chinese officials from the interior who are paid higher salaries, personal subsidies and perks, such as long holidays back to the interior. Recent increases in government wage levels in the western regions to equalise wages across the country and attract talent is likely to increase these pressures further, although at present these pay-rises are officially being met with central government subsidies.

Revenue from mining, both legal and quasi-legal, also finds its way into the pockets of individual officials in the form of cash or goods. This can lead to considerable resentment amongst communities. As the Mining, Minerals and Sustainable Development (MMSD) Project noted:

> *"If politicians or officials siphon mining revenues into their own pocket, local people will reasonably conclude that mining brings them little benefit."*[34]

Interviews with Tibetans in India and Nepal give a strong impression of general distrust of many officials, Tibetan and Chinese, and an assumption (based on experience) that the officials benefit personally from local economic activity, including mining.[35] This reinforces the disparities between the material wealth of the local population and officials. According to a Tibetan nomad from Qinghai, a senior religious teacher made the following pointed remark on the distribution of profits from a local gold mine:

> *"These days when the officials of your county approach, their leather jackets shine, the streets of the county have become swanky. Of course, you people are mining gold! But pity the poor people from the mountain [where the gold is mined]."*

In addition, capital is not necessarily re-invested in the region, but instead flows towards China's coastal areas, where rates of return on investments are generally higher and risks are lower. Capital inflow into Tibet in the form of subsidies is much publicised as exemplifying the benevolence of the state, but little is said of capital outflow. However, Chinese officials assigned to Tibet often see their stay as a sojourn, the main attraction being the high wages and opportunity for making extra income – personal gains that leave with them when they are transferred back to the interior.

34 MMSD, op cit, p.15

35 For an analysis of how officials throughout China have used their positions to enrich themselves see He Qinglian, 'China's Descent into a Quagmire', published in English in three parts in 'The Chinese Economy': vol.33#3, 2000, pp.3-88; Vol.34#2, 2001, pp.3-96; and Vol.34 #4, 2001, pp.3-94

Marginalisation and loss of livelihood

> *"What will be more crucial to Tibetans than an opportunity for new jobs is whether or not the same patterns of change are repeated in the Tibetan heartland as have happened in Haidong, Tsojang and Tsonub. Tibetans are a minority population in all three of these areas."*[36]

Previous chapters have demonstrated that some Tibetan communities have enjoyed benefits from mining, including cash compensation for land use, employment opportunities, infrastructure improvements and wealth generation in their localities. There has also been opportunity for some Tibetans in training and skills development, either within mines or through vocational education.

However, for most Tibetans these largely short-term benefits (which are by no means universal) are outweighed by the long-term costs of mining under current policy and practice. While the state and Chinese industry benefit from resource exploitation, Tibetan communities have to live with the environmental consequences of resource exploitation: the serious damage to the local ecology, primarily through destruction of grasslands and contamination of water sources, as well as through decimation of local wildlife. Another major concern is the continued demographic restructuring of Tibetan areas, which inevitably leads to economic, social and cultural marginalisation.[37]

Despite China's assertions that it is looking to protect both the environment and 'minority nationality' culture, the development of industry on the basis of resource exploitation continues to present a great threat to both. In their critique of development policy, Hu and Wen imply that under current development policy, environmental damage and the 'destruction' of ethnic culture are two of the key negative impacts of industrialisation facing Tibet:

> *"Not only have developments in industrialisation and advancement of a modern civilisation caused the environment to suffer unprecedented damage and pollution, they have also brought about the destruction of the cultural habits and ways of life of ethnic groups [min zu] and the demise of the cultural variety of ethnic groups throughout the world and particularly in developing nations. This has been proved by the history of the development of modern industry and civilisation."*[38]

36 Marshall and Cooke, op cit.
37 See Chapter 3, pp. 84-94
38 Hu and Wen, op cit.

The economic and social marginalisation of local communities, particularly indigenous peoples, and the loss of traditional livelihoods through mining development has been recognised internationally as an important issue. A Control Risk Group analysis of a survey of 51 European oil, gas and mining companies, published in 1997, stated:

> *"[Indigenous people's] may be threatened by population expansion of the dominant group and – often in tandem – threats to the local ecology which make their traditional lifestyles unsustainable".*[39]

Once the traditional livelihoods of Tibetans are undermined through land expropriation and degradation, immigration and marginalisation, and they are no longer able to live in a self-sufficient manner, they become more dependent on the Chinese economy – and as such on China itself. Lifestyles also change, and again because the dominant model is Chinese, this leads to assimilation into the wider Chinese social and cultural model.

Tibetans do not view development, modernisation and change as negative *per se*; however, under the current model, the average Tibetan has no say in how this will take shape. Policies are imposed in a top-down manner, as they are throughout China, and the average Tibetan remains subject to the interests of the Chinese authorities.

Far from having control over their economic futures, China's autonomous areas are obliged to speed up 'economic and cultural construction' in line with central plans. According to China's Regional National Autonomy Law (1984; amended 2001), autonomous areas must "*place the interests of the state as a whole above anything else and make positive efforts to fulfil the tasks assigned by state organs at higher levels*" (Article 7). Autonomous areas have some influence over how economic policy can be implemented, but no influence over what that policy is to be. This is clearly established in Article 25:

> *"Under the guidance of state plans, according to the special characteristics and needs of the local area and to the stipulated guidelines, policies and plans for economic construction, the organs of self-government of national autonomous areas shall independently arrange for and administer local economic development."*

The recent amendments to China's Regional National Autonomy Law, enacted on 28 February 2001, further underline the lack of choice Tibetans have over development. Rather than dealing with issues of autonomy, these amendments serve to bring the law into line with the drive to develop the western regions of China, introducing provisions to increase state control over and facilitate resource exploitation.[40]

39 'No Hiding Place: Business and the politics of pressure', Control Risk Group, 1997
40 See 'National autonomy law revised to support Western Development', TIN News Update, 13 March 2001

Long vs short-term development – impact of mine closures

The level of disruption caused to local communities and economies when mines are closed is an important factor in the sustainability of development based on resource exploitation. Even large-scale mining ventures have a relatively short life span, dependent on the size of exploitable minerals in a particular deposit. Those mines that are not closed for economic or political reasons will eventually close when a deposit's resources are exhausted. A Tibet Daily report listed "*insufficient reserves of natural resources*" as one of the four key problems facing the development of the mining industry in Tibet (along with lack of capital, a high level of credit and debt and low quality of management). The article stated:

> *"Because the objects of exploitation are non-renewable resources, every single mine will have a day when it will close due to the exhaustion of resources, with a direct impact on the work and lives of the mine's employees."*[41]

However, mine closures can often have a much wider impact than the need to find alternative labour for employees, particularly where local economies have become inter-linked with or dependent on mining. This problem is not only an issue in developing countries, but has also been a major issue in developed countries. The fate of the coal mining towns of the UK after the closure of the pits (shut down for economic and political reasons rather than exhaustion of resources), is a clear indicator of the impact on a local economy and population of the problems inherent in resource exploitation when a long-term approach to development is not incorporated into resource extraction plans.[42]

Although most Tibetans remain dependent on animal husbandry and agriculture, mining enterprises have become an important source of income in some areas. While revenue has largely been used for administration, local communities can also be directly affected. For example, Machu county in Kanlho TAP, Gansu province, is considered one of the richest Tibetan nomadic areas, in part due to its gold mining industry. Reports from the area indicate that county officials are already worried about future government income once resources are exhausted. There will also be a significant impact on local people in Nyima township, where the mine is located, because at present the township uses revenues from the mine to pay the taxes of the population.[43] The area has undoubtedly benefited financially in the short-term from

41 'Four Difficult Problems Facing Mining Enterprises in Tibet', Tibet Daily, 12 December 2001

42 The environmental issues associated with mine closure are dealt with in Chapter 5.

43 See pp. 105-107 for more information.

gold mining, but if these benefits are not translated into alternative development they will be limited. With the added factor of environmental degradation, the area could end up adversely affected in the long run.

As the Chinese government pushes for large-scale exploitation in Tibetan areas, the willingness of relevant authorities to implement long-term development plans will be a good indicator of the priorities of the Chinese authorities and the extent to which they are genuinely concerned with the sustainable development of Tibetan areas and enrichment of Tibetan people.

The largest potential impact of mine closures would be on the Chinese settler population of the Tsaidam Basin, where local economies, rather than being transformed, have actually been created by the mining industry. The Chinese workers employed in the extractive industries of Tsonub form the largest concentration of mine workers in any Tibetan area. Although Chinese are not indigenous to the area, official population statistics now list tens of thousands of Chinese settlers as the majority of the population of the mining towns that have been established in the Tsaidam Basin since the 1950s. For example, the urban infrastructure and settlement of Lenghu administrative committee (county-level) was created to exploit the petroleum resources of the area. According to official data, the area's economy does not have an agricultural sector. Without oil, there would no longer be any purpose to settlement of the area and Lenghu's 22,224 employees[44] would need to go elsewhere to find alternative employment.

Gold mining near Lake Manasarovar: trench to supply water, panned for gold **© TIN**

44 Qinghai Statistical Yearbook, 2001

Accountability, participation and international involvement

Sustainable development principles

Over the past decade there has been increasing international focus on sustainable development. This has had a significant impact on the expectations of Western civil society regarding the conduct of mining companies and has led to a broadening of their responsibilities. Although the focus has remained primarily on environmental sustainability, there is increasing pressure to incorporate minority and civil and political rights into codes of conduct, in line with the general global shift towards rights-based development. For example, the charter of the International Council on Minerals and Mining accepts that one of the "*specific challenges which will need to be addressed*" is to promote the health and respect the fundamental human rights of the minerals industry's workers and affected communities.

Accountability and participation are two of the foundations of the sustainable development concept. International standards now recognise the rights of peoples to be consulted on exploration and exploitation of resources in their areas.[45] The increasing importance being attached to the rights of communities in mining development was highlighted by the establishment of the Mining, Minerals and Sustainable Development Project (MMSD) in 2000. The aim of the project was to carry out independent research and consultation into "*how the mining and minerals sector can contribute to the global transition to sustainable development*". MMSD, funded by more than 40 commercial and non-commercial organisations, is a project of the International Institute for Environment and Development (IIED) and was commissioned by the World Business Council for Sustainable Development (WBCSD). The project's report, published in the run-up to the World Summit on Sustainable Development in Johannesburg in August 2002, provides a list of sustainable development principles, including civil and political liberties and cultural autonomy, protection of minority rights, accountability and subsidiarity (decentralisation of decision-making).[46] It gives the following definition of what sustainable development at a local level should mean:

45 For example, ILO Convention (No. 69) concerning Indigenous and Tribal Peoples in Independent Countries (which China has not ratified): Article 15

1. The rights of the peoples concerned to the natural resources pertaining to their lands shall be specially safeguarded. These rights include the right of these peoples to participate in the use, management and conservation of these resources.

2. In cases in which the state retains the ownership of mineral or sub-surface resources or rights to other resources pertaining to lands, governments shall establish or maintain procedures through which they shall consult these peoples, with a view to ascertaining whether and to what degree their interests would be prejudiced, before undertaking or permitting any programmes for the exploration or exploitation of such resource pertaining to their lands. The peoples concerned shall wherever possible participate in the benefits of such activities, shall receive fair compensation for any damages which they may sustain as a result of such activities.

46 See next page

> *"Sustainable development at the local level is about meeting locally defined social, environmental, and economic goals. The effect of interactions between a mine and a local community should add to the physical, financial, human and information resources available to the community. Mineral activities must ensure that the basic rights of the individuals and communities affected are upheld and are not infringed upon. These include the rights to control and use land, to clean water, to safe environment, and to a livelihood; the right to be free from intimidation and violence; and the right to be fairly compensated for loss. The interests of the most vulnerable groups must be protected. There should be equitable distribution of benefits within communities – benefits that are sustainable after the life of a mine."*[47]

Implementation

The difficulties of putting sustainable development principles into practice are considerable and are not limited to the developing world. This is demonstrated by the fact that the national authorities of western liberal democracies, often the driving force behind the formulation of such principles, have themselves sometimes proved unwilling to fully commit to them in the face of other priorities and conflicting interests.

Formal commitments to both environmental protection and sustainable development simply function as window-dressing unless substantive measures are taken to establish mechanisms that will ensure that these principles are carried through in practice. This is the case for international institutions, mining companies (often accused of providing public relations spin and not much more) and governments. Mining companies, investors and donors have to rely upon governments to ensure proper implementation of agreed principles. At present China does not have working mechanisms that ensure accountability and transparency. Moreover, the Chinese authorities are yet to demonstrate a genuine willingness to implement such principles due to overriding economic and political interests.

The Chinese authorities have made an effort to keep abreast of politically correct terminology and have made a formal (if vague) commitment to 'sustainable development' in policy statements and development plans.[48] However, the authorities do not appear to have given any clear definition of what they actually mean by 'sustainable development'. At present, environmental protection and rational use of

46 'Breaking new ground: mining, minerals and sustainable development', MMSD (IIED), 2002, Executive Summary, Table 1. The executive summary and the full text of the report can be downloaded from the IIED website www.iied.org
47 ibid.
48 For example, the Tenth Five-Year Plan uses the terms 'sustainable' and 'rational' development.

resources are likely to be the principal foundations of such a concept in China. The Chinese government is still some way from meeting basic international standards on the protection of civil and political rights and labour rights, and is yet to ratify the UN Convention on Civil and Political Rights or the two core International Labour Organisation (ILO) conventions on forced labour.

Development and policy-making in China is not participatory, but remains imposed from the top down, with little room for consultation. Regional interests in China are determined by state plans and local interests are often marginalised due to a lack of power in decision-making. People are also not provided with the necessary information to make informed decisions about their development future – for example locals in Tibetan often appear to be unaware that a mining project is even to take place until mining teams arrive in the area.

In addition, development principles and policies are inextricably linked with the integrity of the Chinese Communist Party (CCP) and resources are owned by the state. As demonstrated in Chapter 3, local opposition to mining can be construed as "*damaging the construction of the nation*". At present, the political nature of development means that in Tibetan areas it is not possible for opinions heard without fear of reprisal.[49] As the Mining Minerals and Sustainable Development Project report points out:

> *"Ultimately, those who cannot say 'no' to development cannot say 'yes' in any meaningful way either".*

Therefore, what China means by 'sustainable development' may actually be quite different from even the most generally internationally accepted understandings of the concept both specifically within the context of mining and within the context of more general development.

International organisations acting in support of China's development policy in Tibet and other western regions (for example the involvement of the World Bank in promoting private investment in resource exploitation) have their own principles and codes of conduct, which are increasingly moving towards incorporation of minority and human rights. However, in order to abide by such principles it is necessary to see beyond China's formal commitments to sustainable development, poverty alleviation and protection of minority nationality culture. The situation on the ground in Tibetan areas remains outside the framework of much of the current international dialogue regarding sustainable development.

49 See, for example, the World Bank Inspection Panel report on the proposed project to relocate nearly 58,000 predominantly Hui Muslim and Han Chinese farmers from Haidong prefecture in Qinghai, into Dulan county in Tsonub M&TAP. The independent inspection panel concluded that there was a 'climate of fear' amongst the local population, undermining the positive results of the social impact assessments carried out by the Bank.

For example, according to the MMSD project report, it is civil society organisations, including human rights, environmental and community-based groups, policy institutes, churches and charities, who "*ensure a diversity of views are heard in decision-making and are critical as sources of aid and means for its delivery*". However, within Tibetan areas, there is a marked lack of non-governmental and independent civil society organisations able to participate in debate and decision-making. The Chinese authorities need to demonstrate acceptance of the broader principle of independent organisations before the discussion of their inclusion in mineral development is relevant. To give a second example: the principle of protection of worker's rights relies upon a system of accountability, usually implemented by independent trade unions. However, the official All China Federation of Trade Unions (ACFTU) is government controlled. Independent trade unions are not permitted in China and independent trade unionists and labour activists continue to face the risk of imprisonment.

Katarina Tomaševski, academic and UN High Commission for Human Rights Special Rapporteur on the Right to Education (1999/2001), drew attention to the dangers associated with the recent shift in development policy to emphasise partnership with recipients of development aid:

> *"Its positive side is the underlying recognition of the recipient as partner rather than object of development interventions, but its negative side is that this assumption of partnership, that permeates development policy, cloaks inequalities and abuses."*[50]

Despite the above reservations there are some signs of a growing understanding in the international community that Tibetan participation is fundamental to the sustainable development of Tibetan areas. An international seminar, organised by the Chinese central government and the TAR government, was held on 25 and 26 June 2002, with the aim of raising international aid for the TAR as part of the drive to develop the western regions of China. At the symposium UN officials raised the issue of local participation in development. Kerstin Leitner, representative for the United Nations Development Programme (UNDP) in China, stated that Tibetans should be allowed more say in the development of Tibet and the protection of their cultural heritage in order to achieve sustainable and culturally viable development:

> *"A major guarantee for the success of any development process is the active participation and capacity building of the local population. Without their active engagement and co-operation, any development programme will not take root in the Tibetan community and thus will only have a very limited impact."*[51]

50 & 51 See next page

The politics of development

Tibetans' control over their land and resources is limited and fragile. They are not involved in decision-making processes and generally have little ability to hold local authorities accountable. The Chinese Communist Party, which basically determines the economic, social and cultural future of Tibetans, remains unaccountable to them. Therefore, it is hardly surprising that most Tibetans view resource exploitation and other development with suspicion.

Despite the fact that Tibetans have not been the principal beneficiaries of resource exploitation and industrialisation, the Chinese attitude remains paternalistic. Tibetans are expected to be grateful for the 'aid' that they receive from the state and for any development that occurs, whether or not there is a positive or negative impact on their lives and livelihoods. In her report on China's 'Aid-Tibet' projects, Jin Wei, although acknowledging certain failings, states that the projects will have an important impact on the modernisation of Tibet. Moreover, she says, they will "*have a positive influence on the Tibetan masses, heightening their psychological feeling of being members of the big family of various nationalities*" and will play an "*important and irreplaceable role in maintaining Tibet's social stability and development.*"[52]

This points to the underlying politicisation of the development process in Tibetan areas of the PRC. These political aspirations of economic development – to further assimilate Tibetans into the greater Chinese state and to maintain stability – should be clearly understood in any discussion of development or poverty alleviation in Tibetan and other minority nationality areas. Although foreign investors and donors eschew political engagement with recipient states, the politics of development cannot be ignored. Michael Dillon, director of the centre for contemporary Chinese studies at the University of Durham commented on the politicisation of development in an article on the Western Development campaign. He said that under the present circumstances, given the restiveness in Chinese western regions particularly amongst Tibetans and Uighurs, "*economic development can never be merely a neutral device for the alleviation of poverty. It is a conscious political tool, designed to stabilise the western regions.*"[53]

50 (See previous page) Katarina Tomaševski (Professor of International Law and International Relations, University of Lund), 'Minority Rights in Development Aid Policies', Minority Rights Group International, 2000, p.15
51 (See previous page) AFP, 'Greater say for Tibetans urged', SCMP, 26 June 2002
52 Jin Wei, op cit.
53 Michael Dillon, 'China goes west: laudable development or ethnic provocation?', The Analyst, 6 December 2000.

While the state sees integration as positive; many Tibetans view it as negative. Past experience has shown Tibetans that not only will they receive little benefit from further assimilation into the Chinese state, they will also have to live with the negative impacts of increasing marginalisation. Marshall and Cooke argue that as the economy develops according to the current model, Tibetans have generally become increasingly economically and socially irrelevant:

> *"[Tibetans are] drifting backward and downward as economic graphs show the area moving upward. They are obscured as minorities, their languages, tradition and forms of worship experiencing erasure as social, cultural and economic terrain is re-written in favour of newcomers."*[54]

Multilateral institutions like the World Bank and the UN argue that it is better for Tibetans and other 'minority nationalities' if international agencies are involved in mineral exploitation and other development in western China. They state that they are then in a better position to exercise influence over environmental and social standards and that this should lead to greater accountability. This view is shared by those Tibetans who take the pragmatic line that, as China will take Tibetan resources anyway, it is better to find a way to limit the damage and bring at least some benefit to local communities. However, the state is still the implementing agency and outside institutions can only wield limited influence. Any foreign involvement has to accept the basic principles of China's Tibet policy, most notably the validity of the present model of industrialisation based on resource exploitation and the state ownership and control of resources. If mining Tibet is to contribute to sustainable development for Tibetans, the real challenge is to progress from a situation in which the choice for Tibetans is between the better of two evils to a situation of positive benefit, in which Tibetans are central to the development of their land and resources, rather than being left on the fringes.

54 Marshall and Cooke, op cit.

Hillside behind Nubzur monastery, Serthar (Ch: Seda) county, Kardze (Ch: Ganzi) TAP, mined for gold © TIN

Gold mining near Lake Manasarovar: water from Ganga Chu pumped into trench © TIN

Conclusion

The Chinese authorities have prioritised the exploitation of Tibet's mineral reserves in their efforts to meet a growing national demand and to maximise the potential economic value of Tibet to the Chinese state. To date there has been relatively little foreign investment into China's minerals industry, but international agencies, mining companies and Western academics are already involved in opening up possibilities for future mining development in China's western regions. While official Chinese reports on the potential of mining Tibet are exaggerated and understate the many obstacles to development, there is little doubt that the state continues to see Tibet as an important provider of raw materials for Chinese industry.

The mining industry is growing rapidly in those Tibetan areas that already have established industrial bases, settlements and transport infrastructure, most notably in the Tsaidam Basin and north-east Qinghai. At the same time known reserves of key mineral commodities such as copper, previously unfeasible to exploit, are becoming increasingly accessible. Plans to exploit these reserves and step up geological exploration are opening up new areas to mining development. If implemented, plans to clamp down on small-scale and artisanal mining, and to move towards more organised and rational development, could limit further environmental degradation in existing mining areas. But the reach of the mining industry is extending to areas of the Tibetan plateau that have, until recently, been shielded by their remoteness. This bodes ominously for the future of Tibet's fragile eco-system, particularly given the poor environmental standards known to exist in similar enterprises elsewhere in Tibet and China.

The new railway from Golmud in Qinghai to Lhasa in the Tibet Autonomous Region (TAR), due to be completed by the end of 2007, could prove to be the most important development in altering the economics of ore exploitation in Tibet. Sparsely populated areas along the route of the railway, previously inaccessible, will be opened up to mining and new urban settlements are likely to be created, similar to the mining towns in the Tsaidam basin in Qinghai that are predominantly populated by Chinese settlers. The railway will also facilitate more effective transportation of raw materials from central Tibet to China, thus more economical exploitation of deposits in the TAR, and the in-migration of greater numbers of China's rural poor in search of work.

Concerns are being expressed even within China as to whether resource exploitation can provide a basis for sustained development of Tibetan areas. Current policy appears to be increasing rather than decreasing the local authorities' reliance on subsidies, while investment has been limited in those sectors of the economy that have the potential to generate indigenous, sustainable development at the grass-roots level. Given that the Chinese state asserts ownership of Tibet's minerals and control over economic policy, 'autonomous' Tibetan areas are left with little opportunity to benefit from their richest resources and have no significant voice in their economic futures. In practice, China's 'minority nationality' autonomous areas appear to have less autonomy than the economic zones of China's east coast, where both officials and business leaders wield considerable influence due to their economic power and political connections.

The current model of development in Tibet appears to be enhancing disparities between rich and poor, and between Tibetans and Chinese, reflecting a similar pattern in other developing countries where mining has become the basis of local economic growth. Tibetans generally remain on the fringes of an industry that threatens their traditional livelihoods, but at present offers little in the way of long-term alternatives. The majority of Tibet's population are farmers and nomads who have close religious, cultural and economic ties to the land and rely upon it for their livelihoods. Mining projects often pose a threat to this religious and economic relationship with the land.

The Chinese Communist Party's main priorities for national development are to sustain China's economic growth while preventing political disintegration. This means exploiting the resources of the west to support increasing prosperity driven by industrialisation and urbanisation in the rest of the country, and drawing restive 'minority nationality' areas into the greater Chinese economic, cultural and social model. However, even Chinese academics are now pointing to flaws in central development policy and are arguing for the urgent need to move towards a model of development more appropriate to Tibet and its fragile environment and culture – reasoning that this is in the best interests not only of Tibet, but of China. Tibet is not simply a treasure-house of minerals. It is a vital, but fragile, eco-system, the protection of which is essential not only to the lives of its indigenous population, but to China, Asia and beyond.

Mining Tibet – Acronyms

BP	British Petroleum
CCP	Chinese Communist Party
CTC	Canada Tibet Committee
EIA	Environmental Impact Assessment
EPB	Environmental Protection Bureau
FBIS	Foreign Broadcast Information Service
FEER	Far Eastern Economic Review
ICMM	International Council on Minerals and Mining
IIED	International Institute for Environment and Development
ILO	International Labour Organisation (UN body)
IMM	Institute of Mining and Metallurgy
M&TAP	Mongolian and Tibetan Autonomous Prefecture
MLR	Ministry of Land and Resources
MMSD	Mining, Minerals and Sustainable Development
MPM	Monopolistic Purchase and Management
MRL	Mineral Resources Law
NPC	National People's Congress
PBC	People's Bank of China
PRC	People's Republic of China
RMB	Renminbi (People's currency)
SEPA	State Environmental Protection Agency
SMI	Sino Mining International
SOE(s)	State-owned Enterprise(s)
SWB	Summary of World Broadcasts (BBC Monitoring)
T&HAC	Tu and Hui Autonomous County
T&QAP	Tibetan and Qiang Autonomous Prefecture
TAC	Tibetan Autonomous County
TAP	Tibetan Autonomous Prefecture
TAR	Tibet Autonomous Region
TVE(s)	Township and Village Enterprise(s)
UNDP	United Nations Development Programme
UNFC	United Nations Framework Classification
UNHRC	United Nations Human Rights Commission
USGS	United States Geological Survey
WBCSD	World Business Council for Sustainable Development
WBG	World Bank Group
WGC	World Gold Council
WTO	World Trade Organisation
XUAR	Xinjiang Uighur Autonomous Region
YAP	Yi Autonomous Prefecture

Mining Tibet – Place name translations (TAR)

TIBET AUTONOMOUS REGION		TIBETAN	CHINESE (pin yin)	CHINESE (simplified)
		བོད་རང་སྐྱོང་ལྗོངས་	Xizang Zizhiqu	西藏自治区
MAP REF.	COUNTY TOWN			
Chamdo Prefecture		ཆབ་མདོ་ས་ཁུལ་	*Changdu Zhou*	昌都州
	Chamdo	ཆབ་མདོ་	Changdu	昌都
D1	Dengchen	སྟེང་ཆེན་	Dingqing	丁青
D2	Dragyab	བྲག་གཡབ་	Chaya	察雅
D3	Jomda	འཇོ་མདའ་	Jiangda	江达
D4	Riwoche	རི་བོ་ཆེ་	Leiwuqi	类乌齐
*Lhasa Municipality**		ལྷ་ས་གྲོང་ཁྱེར་ཁུལ་	*Lasa Shi*	拉萨市
	Lhasa City	ལྷ་ས་གྲོང་ཁྱེར་	Lasa Chengguanqu	拉萨城关区
C1	Maldrogongkar	མལ་གྲོ་གུང་དཀར་	Mozhugongka	墨竹工卡
C2	Nyemo	སྙེ་མོ་	Nimu	尼木

	Lhoka Prefecture	ལྷོ་ཁ་ས་ཁུལ་	*Shannan Zhou*	山南州
	Nedong	སྣེ་གདོང་	Naidong	乃东
	Chusum	ཆུ་གསུམ་	Qusong	曲松
	Nagchu Prefecture	ནག་ཆུ་	*Naqu Zhou*	那曲州
	Nagchu	ནག་ཆུ་	Naqu	那曲
B1	Amdo	ཨ་མདོ་	Anduo	安多
B2	Shantsa	ཤན་རྩ་	Shenzha	申扎
	Ngari Prefecture	མངའ་རིས་ས་ཁུལ་	*Ali Zhou*	阿里州
	Gar	སྒར་	Ge'er	噶尔
	Nyingtri Prefecture †	ཉིང་ཁྲི་	*Linzhi*	林芝地区
	Nyingtri	ཉིང་ཁྲི་	Linzhi	林芝
	Shigatse Prefecture	གཞིས་ཀ་རྩེ་ས་ཁུལ་	*Rigaze Zhou*	日喀则州
	Shigatse	གཞིས་ཀ་རྩེ་	Rigaze	日喀则
A1	Zhethongmon	བཞད་མཐོང་སྨོན་	Xietongmen	谢通门

* Lhasa Municipality operates at prefectural level and covers seven counties, including Lhasa City.

† Also known as Kongpo Prefecture ཀོང་པོ་ས་ཁུལ་ *Gongbu Diqu* 共布地区

Mining Tibet – Place name translations (Qinghai)

	MAP REF.	COUNTY TOWN	TIBETAN	CHINESE (pin yin)	CHINESE (simplified)
QINGHAI PROVINCE			མཚོ་སྔོན་ཞིང་ཆེན་	Qinghai Sheng	青海省
Golog TAP			མགོ་ལོག་བོད་རིགས་རང་སྐྱོང་ས་ཁུལ་	*Guoluo Zangzu Zizhi Zhou*	果洛藏族自治州
		Machen	རྨ་ཆེན་	Maqin	玛沁
	G1	Pema	པད་མ་	Banma	班玛
Malho TAP			རྨ་ལྷོ་བོད་རིགས་རང་སྐྱོང་ས་ཁུལ་	*Huangnan Zangzu Zizhi Zhou*	黄南藏族自治州
		Rebgong	རེབ་གོང་	Tongren	同仁
Tsoshar Prefecture			མཚོ་ཤར་ས་ཁུལ་	*Haidong Diqu*	海东地区
		Tsongkhakhar	ཙོང་ཁ་མཁར་	Ping'an	平安
Tsojang TAP			མཚོ་བྱང་བོད་རིགས་རང་སྐྱོང་ས་ཁུལ་	*Haibei Zangzu Zizhi Zhou*	海北藏族自治州
	H1	**Chilen/Dola**	ཆི་ལན་ / མདོ་ལ་	Qilian	祁连

Tsolho TAP		མཚོ་ལྷོ་བོད་རིགས་རང་སྐྱོང་ས་ཁུལ་	*Hainan Zangzu Zizhi Zhou*	海南藏族自治州
	Chabcha	ཆབ་ཆ་	Gonghe	共和
Tsonub M & TAP		མཚོ་ནུབ་བོད་སོག་མི་ རིགས་རང་སྐྱོང་ས་ཁུལ་	*Haixi Mengguzu he Zangzu Zizhi Zhou*	海西蒙古族和 藏族自治州
	Terlenkha Munic.	གཏེར་ལེན་ཁ་གྲོང་སྡེ་ཁུལ་	Delingha Shi	德令哈市
	Golmud (Nagormo) Munic.*	ན་གོར་མོ་ གོར་མོ་གྲོང་སྡེ་ཁུལ་	Ge'ermu Shi	格尔木市
F1	Tulan	ཏུའུ་ལན་	Dulan	都兰
F2	Wulan	ཝུ་ལན་	Wulan	乌兰
Siling Municipality		ཟི་ལིང་གྲོང་ཁྱེར་ཁུལ་	*Xining Shi*	西宁市
	Siling	ཟི་ལིང་	Xining	西宁
J1	Kumbum	སྐུ་འབུམ་	Huangzhong	湟中
Yushu TAP		ཡུ་ཤུ་བོད་རིགས་རང་སྐྱོང་ས་ཁུལ་	*Yushu Zangzu Zizhi Zhou*	玉树藏族自治州
	Kyegudo	སྐྱེ་རྒུ་མདོ་	Yushu/Jiegu	玉树/结古
E1	Chumarleb	ཆུ་དམར་ལེབ་	Qumalai	曲麻莱
E2	Dritoe	འབྲི་སྟོད་	Zhiduoi	治多
E3	Tridu	ཁྲི་འདུ་	Chengduo	称多

* Both Terlenkha and Golmud are administered by Tsonub M & TAP in order to optimise resource extraction.

Mining Tibet – Place name translations (Sichuan)

		TIBETAN	CHINESE (pin yin)	CHINESE (simplified)
SICHUAN PROVINCE		ཟི་ཁྲོན་ཞིང་ཆེན་	Sichuan Sheng	四川省
MAP REF.	COUNTY TOWN			
Kardze TAP		དཀར་མཛེས་བོད་རིགས་རང་སྐྱོང་ས་ཁུལ་	*Ganzi Zangzu Zizhi Zhou*	甘孜藏族自治州
	Dartsedo	དར་རྩེ་མདོ་	Kangding	康定
K1	Bathang	སྨར་ཁམས་	Ma'erkang	巴塘
K2	Chagzam	ལྕགས་ཟམ་	Luding	泸定
K3	Dabpa	འདབ་པ་	Daocheng	稻城
K4	Dawu	རྟའུ་	Daofu	道浮
K5	Dege	སྡེ་དགེ་	Dege	德格
K6	Gyezur/Gyazil	བརྒྱད་ཟུར་	Jiulong	九龙
K7	Kardze	དཀར་མཛེས་	Ganzi	甘孜
K8	Lithang	ལི་ཐང་	Litang	理塘
K9	Palyul	དཔལ་ཡུལ་	Baiyu	白玉
K10	Serthar	གསེར་ཐར་	Seda	色达

Liangshan YAP		ལི་ཡན་ཤན་དབྱི་མི་རིགས་རང་སྐྱོང་ས་ཁུལ་	*Liangshan Yizu Zizhi Zhou*	凉山彝族自治州
	Mili Tibetan AC*	མི་ལི་བོད་རིགས་རང་སྐྱོང་ས་ཁུལ་	Muli Zangzu Zizhi Xian	木里藏族自治县
Ngaba T&QAP		ཛ་འཆང་བོད་རིགས་དང་འཆང་རིགས་རང་སྐྱོང་ས་ཁུལ་	*Aba Qiangzu he Zangzu Zizhi Zhou*	阿坝羌族和藏族自治州
	Barkham	འབར་ཁམས་	Batang	马尔康
L1	Dzamthang	འཛམ་ཐང་	Rangtang	壤塘
L2	Dzoege	མཛོད་དགེ་	Ruo'ergai	若尔盖
L3	Kakhog	ཀ་ཁོག་	Hongyuan	红原
L4	Namphing	ནམ་ཕིག	Jiuzhaigou	九寨沟
L5	Zungchu	ཟུང་ཆུ་	Songpan	松潘

* Mili Tibetan Autonomous County lies within Liangshan Yi Autonomous Prefecture.

Mining Tibet – Place name translations (Gansu)

	MAP REF	COUNTY TOWN	TIBETAN	CHINESE (pin yin)	CHINESE (simplified)
GANSU PROVINCE			ཀན་སུ་ཞིང་ཆེན་	Gansu Sheng	甘肃省
Kanlho TAP			ཀན་ལྷོ་བོད་རིགས་རང་སྐྱོང་ས་ཁུལ་	*Gannan Zangzu Zizhi Zhou*	甘南藏族自治州
		Tsoe	གཙོས་	Hezuo	合作
	M1	Luchu	ཀླུ་ཆུ་	Luqu	碌曲
	M2	Machu	རྨ་ཆུ་	Maqu	玛曲
	M3	Thewo	ཐེ་བོ་	Diebu	迭部
Wuwei Prefecture			ཝུའུ་ཝེ་ས་ཁུལ་	*Wuwei Zhou*	武威州
		Pari TAC*	དཔའ་རིས་བོད་རིགས་རང་སྐྱོང་ས་ཁུལ་	Tianzhu Zangzu Zizhi Xian	天祝藏族自治县

* Pari Tibetan Autonomous County lies within Wuwei Prefecture, which has no ethnically defined autonomous status.

Mining Tibet – Place name translations (Yunnan)

		TIBETAN	CHINESE (pin yin)	CHINESE (simplified)
YUNNAN PROVINCE		ཡུན་ནན་ཞིང་ཆེན་	Yunnan Sheng	云南省
MAP REF	COUNTY TOWN			
Dechen TAP		བདེ་ཆེན་བོད་རིགས་རང་སྐྱོང་ས་ཁུལ་	*Diqing Zangzu Zizhi Zhou*	迪庆藏族自治州
	Gyalthang †	རྒྱལ་ཐང་	Zhongdian	中甸
N1	Dechen	བདེ་ཆེན་	Deqin	德钦

† Gyalthang's name was formally changed to Shangri-la in 2001 to promote tourism in the area.

Mining Tibet Orientation Map Key

	Jangthang (northern steppe)
	Tsaidam Basin
	Hoh Xil nature reserve
	Yulong copper belt
	West Qinling gold belt
	3 rivers' sources district
	Disputed regions
	Provincial capital
NGARI	Prefecture
Gar	Prefectural capital
• K6	County town
	Provincial boundary
	Prefectural boundary
315	National highway
	River – lake
	Railway

Key – Rivers and Lakes

r1	River Sutlej	r6	Lake Yamdrok
r2	River Indus	r7	Nu (Salween) River
r3	Lake Rakas Tal	r8	Lancang (Mekong) River
r4	Lake Manasarovar	r9	Jinsha (Yangtze) River
r5	Yarlung Tsangpo (Brahmaputra) River	r10	Machu (Yellow) River
		r11	Lake Qinghai